P₃

The Atomic Nucleus

The Atomic Nucleus

J. M. Reid

Second edition

Manchester University Press

Copyright © J. M. Reid 1972, 1984
First published by Penguin Books Ltd 1972
in the Penguin Library of Physical Sciences
New edition, with corrections and additional chapter,
published 1984 by
Manchester University Press
Oxford Road, Manchester M13 9PL, U.K.
51 Washington Street, Dover, N.H. 07382, U.S.A.

British Library cataloguing in publication data
Reid, J.M.
    The atomic nucleus — 2nd ed.
    1. Nuclear physics
    I. Title
    539.7'23   QC776

Library of Congress cataloging in publication data
Reid, J.M. (John McArthur)
    The atomic nucleus.
    Bibliography: p. 271
    Includes index.
    1. Nuclear physics.   I. Title.
    QC776.R45   1984   539.7   84-7931

ISBN 0 7190-0978-2

Printed in Great Britain by
Unwin Brothers Limited, The Gresham Press, Old Woking, Surrey

Contents

Editorial Foreword

For many years, now, the teaching of physics at the first-degree level has posed a problem of organization and selection of material of ever-increasing difficulty. From the teacher's point of view, to pay scant attention to the groundwork is patently to court disaster; from the student's, to be denied the excitement of a journey to the frontiers of knowledge is to be denied his birthright. The remedy is not easy to come by. Certainly, the physics section of the Penguin Library of Physical Sciences does not claim to provide any ready-made solution of the problem. What it is designed to do, instead, is to bring together a collection of compact texts, written by teachers of wide experience, around which undergraduate courses of a 'modern', even of an adventurous, character may be built.

The texts are organized generally at three levels of treatment, corresponding to the three years of an honours curriculum, but there is nothing sacrosanct in this classification. Very probably, most teachers will regard all the first-year topics as obligatory in any course, but, in respect of the others, many patterns of interweaving may commend themselves, and prove equally valid in practice. The list of projected third-year titles is necessarily the longest of the three, and the invitation to discriminating choice is wider, but even here care has been taken to avoid, as far as possible, the post-graduate monograph. The series as a whole (some five first-year, six second-year and fourteen third-year titles) is directed primarily to the undergraduate; it is designed to help the teacher to resist the temptation to overload his course, either with the outmoded legacies of the nineteenth century, or with the more speculative digressions of the twentieth. It is expository, only: it does not attempt to provide either student or teacher with exercises for his tutorial classes, or with mass-produced questions for examinations. Important as this provision may be, responsibility for it must surely lie ultimately with the teacher: he alone knows the precise needs of his students – as they change from year to year.

Within the broad framework of the series, individual authors have rightly regarded themselves as free to adopt a personal approach to the choice and presentation of subject matter. To impose a rigid conformity on a writer is to dull the impact of the written word. This general licence has been extended even to the matter of units. There is much to be said, in theory, in favour of a single system of units of measurement – and it has not been overlooked that national policy in advanced countries is moving rapidly towards uniformity under the *Système International* (SI units) – but fluency in the use of many systems is not to be despised: indeed, its acquisition may further, rather than retard, the physicist's education.

A general editor's foreword, almost by definition, is first written when the series for which he is responsible is more nearly complete in his imagination (or the publisher's) than as a row of books on his bookshelf. As these words are penned, that is the nature of the relevant situation: hope has inspired the present tense, in what has just been written, when the future would have been the more realistic. Optimism is the one attitude that a general editor must never disown!

N. Feather

Preface

This book is intended for final-year honours-degree students specializing in physics. It is assumed that they have previously studied the elementary nuclear properties. They should have encountered the concept of the Rutherford–Bohr atom, have a qualitative picture of the construction of nuclei from protons and neutrons, and have been introduced to the basic facts of radioactivity. These topics are covered from first principles in this text, but necessarily the elementary facts are very quickly passed over to permit a development of less elementary aspects. It is also assumed that a prior or parallel study of quantum mechanics is included in the course.

The volume of experimental information on nuclear properties, and the extent of relevant nuclear theory, is such as to necessitate a high degree of selection in a treatment such as the present. The selection was undertaken with two aims in view; firstly to provide an account of the basic nuclear properties of size, electric charge, mass, magnetic and electric moments, and secondly to give an account of currently useful nuclear models. A brief description of well-established experimental techniques is provided at the appropriate places, and a more extended, but still of course introductory, discussion of recent developments such as solid-state detectors and the Doppler-shift technique for the measurement of short-lived excited states is included. The selection of topics will be found to reflect the growing interest over the last decade in 'collective' properties of nuclear matter, and also the more recent refocusing of interest on heavy nuclei. The latter has stemmed from the discovery of shape isomerism and the speculations on the existence of 'superheavy' nuclei. The physical basis on which these developments rest, and the reasons for the current general interest in them, is given in some detail. From this introduction the student should be able to see the direction in which the subject of nuclear physics is currently moving and, if proceeding to postgraduate work in this field, it should enable him to reach a starting-off point for the study of specialist monographs and research publications.

It is recognized that the material here presented is probably much more than can be covered in the time available for nuclear physics in a balanced honours course. It will however be found that the chapters are more or less self-contained, and it should be relatively simple to further select the topics to be covered in a particular syllabus. There should then be sufficient additional material to allow the better student to reach beyond the immediate needs of the syllabus, and also to impress on all the students how selective the coverage has been.

The selection of material and the presentation are entirely the responsibility of the author. Acknowledgement is gratefully made of the author's debt to Alastair MacLeod, who read the original manuscript, for his most useful comments. Kenneth Ledingham was kind enough to advise on the general subject of weak interactions. The author also wishes to express his thanks to N. Feather who, in his role as general editor of this series, invited and encouraged this production, and who, in his role as nuclear scientist, constructively criticized the material dealing with the early development of the subject.

Finally, the author wishes to acknowledge his debt to P. I. Dee under whose general direction he has been privileged to research and teach in the field of nuclear physics. Professor Dee's continuing interest in the presentation of this topic to undergraduates has been a great stimulation and encouragement.

Chapter 1
Introduction

1.1 Historical introduction

By the end of the nineteenth century, the successful development of chemical science had firmly established the concepts of *molecule* and *atom*. A sample of a chemical compound, it was believed, could be divided into parts, the parts having the same chemical behaviour as the original sample. Subdivision could continue without change in the chemical behaviour until the resulting specimen consisted of a single molecule, but if subdivision continued beyond this stage the chemical behaviour was no longer unaffected. Molecules were believed to be assemblies of a comparatively small number of atoms of which about ninety different types were thought to exist, each type corresponding to a chemical *element*. To take a simple example, the molecule of water was considered to be built from three atoms, two corresponding to the element hydrogen and one to the element oxygen. Its behaviour was that of water (or water vapour) unless it was dissociated into its component atoms, in which case it behaved as a mixture of the gaseous elements oxygen and hydrogen. Each type of atom was believed to be of different structure, and this structure in some way determined the chemical behaviour of the element with which the atom was associated.

Shortly before the beginning of the present century, there were indications that, just as molecules had internal structure in the sense of being constructed from atoms, so atoms themselves had structure and were built from more fundamental entities. The discovery of the electron (J. J. Thomson, 1897), which was found to be associated with a wide range of materials of different chemical behaviour, and the realization that its mass was very much less than that of the lightest atom, provided an early example of a possible 'subatomic' particle. At this time too, the study of radioactivity, a phenomenon Becquerel had stumbled upon, in 1896, in an investigation into the fluorescence of uranium salts following the then recent discovery of X-rays, provided examples of parts of atoms being ejected in some form of internal reorganization. Various models conferring internal structure on the atom were current in the first decade of this century. Rutherford in 1911, on the basis of alpha-particle scattering (which is discussed in detail in Chapter 3), proposed a new model. This model, with features added by Bohr in 1913, remains the basis of present-day theory.

1.2   The Rutherford–Bohr atom

According to the currently accepted picture, the atom is a comparatively diffuse empty structure, having a concentrated core or *nucleus*. Orbiting around the nucleus at distances very much greater than nuclear dimensions are electrons. The dimensions of the outermost electron orbits set the size of the atom, as strong repulsive forces operate when atoms come so close together that their electron orbits overlap. The mass of the electrons represents a very small fraction of the total mass of the atom. They are held in their orbits by the Coulomb force of attraction between their negative electric charge and a positive charge which is assumed to reside in the nucleus. At distances greater than atomic dimensions, the positive charge in the nucleus will be 'screened' by the negatively charged orbiting electrons. If the screening is not complete, that is, the charge in the nucleus is not completely compensated by the charge of the electrons, then additional electrons will be collected from the surroundings until the neutralization is exact.

Moseley's work (1913) on X-rays enabled the chemical elements to be placed precisely in order of increasing frequency of characteristic X-radiation. This order also gave the best fit of the elements into the periodic table, which had been developed as a means of revealing the pattern of chemical behaviour in the progression from lighter to heavier elements. The position of an element in this order fixes its *atomic number* (or *charge number*) which is denoted by Z. This number ranges from 1 for hydrogen to 92 for uranium, the highest value for a naturally occurring element (see Appendix E). The simple hypothesis that the number of orbiting electrons in an atom may be equated with Z turns out to be tenable. This hypothesis demands that the nuclear charge be $+ Ze$, where $-e$ is the electric charge on an electron. By alpha-particle-scattering experiments Chadwick (1920) was able to measure the charges on the nuclei of platinum, silver and copper. The values found experimentally were in good agreement with $+ Ze$ in each case.

1.3   The mass of the atom

Appeal to Avogadro's hypothesis, that 'equal volumes of all gases under the same conditions of temperature and pressure contain the same number of molecules', enables the relative weights (and hence masses) of molecules to be obtained from a comparison of gas densities. From a knowledge of molecular composition, in terms of constituent atoms, relative atomic weights may then be deduced. Standard atomic weights, as used in chemistry, are obtained from these relative values by defining the atomic weight of oxygen to be 16.

If the mass of an atom in absolute terms is required, then the total number of molecules in a given volume must be known. Usually the volume chosen is the gramme-molecular volume, which is 22·4 l. At 0 °C and one atmosphere pressure the number of molecules in this volume is known as *Avogadro's constant*. By several independent experimental methods estimates have been made of this number and the results are all in good agreement with the value $6·023 \times 10^{23}$.

Since 22·4 l of hydrogen at 0 °C and one atmosphere pressure weigh 2 g, and since there are two hydrogen atoms per molecule, it follows that the mass of the hydrogen atom is $1·67 \times 10^{-24}$ g. Numerically this is the reciprocal of Avogadro's constant.

Some, but not all, chemical atomic masses are close to integral multiples of the hydrogen mass. This led to the very early suggestion, made by Prout in 1815, that the atoms of all elements are combinations of hydrogen atoms. Unfortunately the existence of elements whose chemical atomic masses fell half-way between integral multiples of the hydrogen atomic mass constituted an insuperable objection to this simple hypothesis. However, the situation was basically altered by the discovery of the existence of *isotopes*. This discovery arose in the chemical investigation (Soddy, 1906–13) of the products of radioactive decay. To take one instance, lead was produced by the decay of radium F. Lead was also produced by the decay of thorium C'. The two forms of lead were chemically indistinguishable from each other (and from ordinary lead) but one was found to have atomic mass 206, the other 208. These were then said to constitute two different isotopes of lead. The work of J. J. Thomson (1913) on positive ions revealed that there were two distinct isotopes of different mass present also in neon. It is now known that very many of the chemical elements occurring in their natural form, including hydrogen and oxygen, consist of a mixture of isotopes (see Appendix A). The chemical atomic mass, in the case of an element which has two or more stable isotopes, is an average value which depends on the relative abundance of the isotopes. If now we consider the isotopic masses instead of the chemical atomic masses, then, as will be discussed in detail later, a mass scale can be defined on which all isotopic masses have a value lying close to an integral number. For any particular isotope this integral number is called the *mass number* and is denoted by A.

1.4   The size of the atom

Since the atomic weight of oxygen is sixteen, the weight of a water molecule is approximately eighteen times the weight of a hydrogen atom, i.e. it is about 30×10^{-24} g. In 1 cm^3 of water, which weighs 1 g, there are thus $10^{24}/30$ molecules. Each molecule therefore occupies 30×10^{-30} m^3, and so the molecular linear dimension must be approximately 3×10^{-10} m. By simple arguments of this kind the atomic radius, assuming the atom to be spherical, is deduced to be of the order of 10^{-10} m. Alpha-particle-scattering experiments show the uranium nucleus on the other hand to have a radius smaller than 3×10^{-14} m. We thus see the extent to which the atom is 'a diffuse empty structure'. The radius of the nucleus is about 10^{-4} times the radius of the atom. If the nucleus is scaled to the size of the earth (6400 km radius) then the atom would extend to about 64×10^6 km, that is, to a distance comparable to the sun's distance from the Earth.

O 15·9994   8   σ < 0·0002

N 14·0067   7   σ 1·9

C 12·01115 σ 0·0034	C 9 0·13s β+ 3·5 (p 9·3, 12·3)	C 10 19s β+ 1·9,·· γ 0·72, 1·0| E 3·6

6

B 10·811 σ 759	B 8 0·78s β+ 1·4 (2α3) E 18	B 9 ⩾3 × 10⁻¹ ṗ, (2α) 9·01333

5

Be 9·0122 σ 0·009	Be 6 ≳4 × 10⁻²¹s p, α,⁵Li 6·0198	Be 7 3/- 53d ε γ 0·48 σnp 5400 E 0·86	Be 8 ~3 × 10⁻¹ 2α 0·09 8·00531

4

Li 6·939 σ 71	Li 5 3/- ~ 10⁻²¹s p, α 5·0125	Li 6 1+ 7·42 σnα 950 σ 0·045 6·01513	Li 7 3/- 92·58 σ 0·036 7·01601

3

He 4·0026 σ 0·007	He 3 1/+ 0·00013 σnp 5327 3·01603	He 4 ~ 100 σ 0 4·00260	He 5 3/- 2 × 10⁻²¹s n, α 5·0123	He 6 0·81s β- 3·51 no γ E 3·51

2

H 1·00797 σ 0·33	H 1 1/+ 99·985 σ 0·33 1·007825	H 2 1+ 0·015 σ 0·0005 2·01410	H 3 1/+ 12·26y β- 0·0181 no γ σ < 0·000007 E 0·0181		H 5? 0·1s β- > 15

1

3   4

n 1 1/+ 12m β- 0·78 E 0·78 1·008665

0

0   1   2

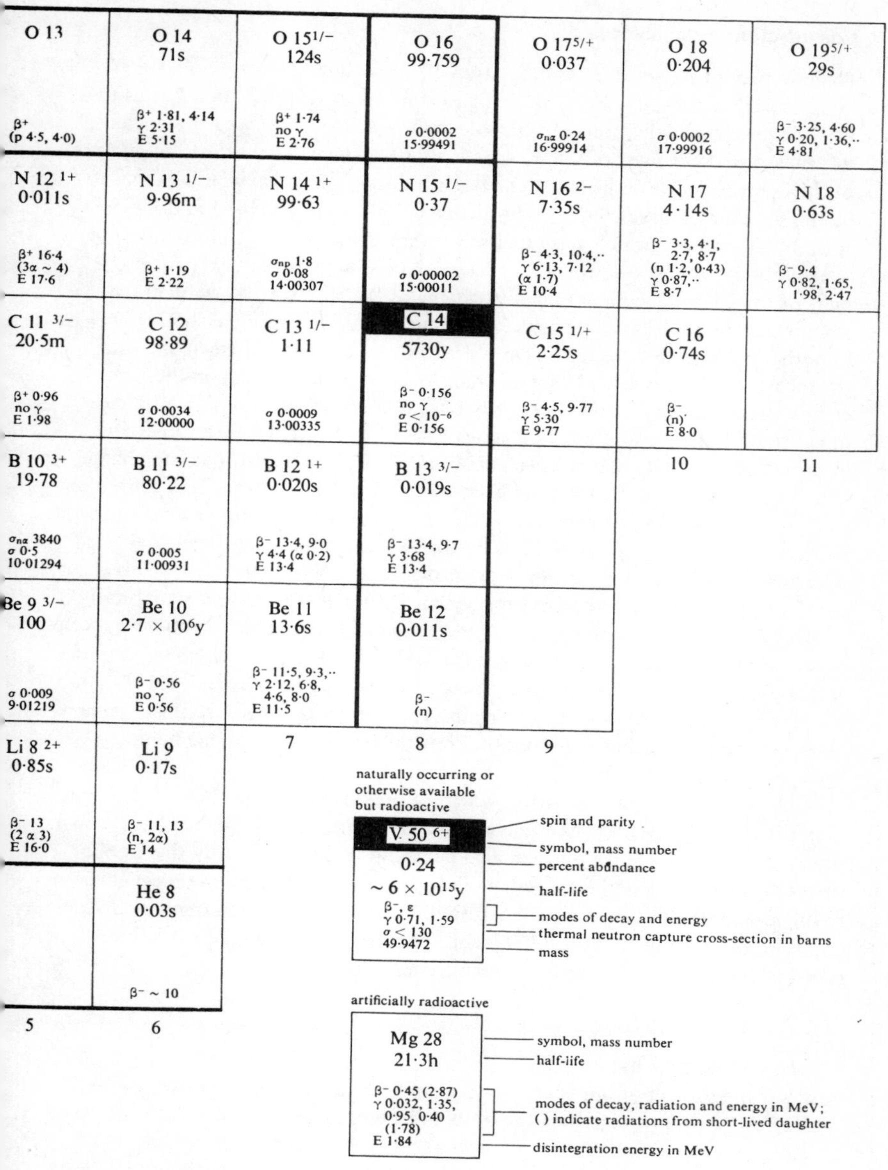

N = 5	N = 6	N = 7	N = 8	N = 9	N = 10	N = 11
O 13 β+ (p 4·5, 4·0)	O 14 71s β+ 1·81, 4·14 γ 2·31 E 5·15	O 15 1/− 124s β+ 1·74 no γ E 2·76	O 16 99·759 σ 0·0002 15·99491	O 17 5/+ 0·037 σ$_{na}$ 0·24 16·99914	O 18 0·204 σ 0·0002 17·99916	O 19 5/+ 29s β− 3·25, 4·60 γ 0·20, 1·36,·· E 4·81
N 12 1+ 0·011s β+ 16·4 (3α ~ 4) E 17·6	N 13 1/− 9·96m β+ 1·19 E 2·22	N 14 1+ 99·63 σ$_{np}$ 1·8 σ 0·08 14·00307	N 15 1/− 0·37 σ 0·00002 15·00011	N 16 2− 7·35s β− 4·3, 10·4,·· γ 6·13, 7·12 (α 1·7) E 10·4	N 17 4·14s β− 3·3, 4·1, 2·7, 8·7 (n 1·2, 0·43) γ 0·87,·· E 8·7	N 18 0·63s β− 9·4 γ 0·82, 1·65, 1·98, 2·47
C 11 3/− 20·5m β+ 0·96 no γ E 1·98	C 12 98·89 σ 0·0034 12·00000	C 13 1/− 1·11 σ 0·0009 13·00335	**C 14** 5730y β− 0·156 no γ σ $< 10^{-6}$ E 0·156	C 15 1/+ 2·25s β− 4·5, 9·77 γ 5·30 E 9·77	C 16 0·74s β− (n) E 8·0	
B 10 3+ 19·78 σ$_{na}$ 3840 σ 0·5 10·01294	B 11 3/− 80·22 σ 0·005 11·00931	B 12 1+ 0·020s β− 13·4, 9·0 γ 4·4 (α 0·2) E 13·4	B 13 3/− 0·019s β− 13·4, 9·7 γ 3·68 E 13·4		10	11
Be 9 3/− 100 σ 0·009 9·01219	Be 10 $2·7 \times 10^6$ y β− 0·56 no γ E 0·56	Be 11 13·6s β− 11·5, 9·3,·· γ 2·12, 6·8, 4·6, 8·0 E 11·5	Be 12 0·011s β− (n)			
Li 8 2+ 0·85s β− 13 (2 α 3) E 16·0	Li 9 0·17s β− 11, 13 (n, 2α) E 14	7	8	9		
5	He 8 0·03s β− ~ 10					
	6					

naturally occurring or otherwise available but radioactive

V 50 6+
0·24
$\sim 6 \times 10^{15}$ y
β−, ε
γ 0·71, 1·59
σ < 130
49·9472

— spin and parity
— symbol, mass number
— percent abundance
— half-life
— modes of decay and energy
— thermal neutron capture cross-section in barns
— mass

artificially radioactive

Mg 28
21·3h

β− 0·45 (2·87)
γ 0·032, 1·35,
0·95, 0·40
(1·78)
E 1·84

— symbol, mass number
— half-life
— modes of decay, radiation and energy in MeV; ( ) indicate radiations from short-lived daughter
— disintegration energy in MeV

Figure 1 Part of the *Chart of the Nuclides* prepared (1966) by D. T. Goldman and J. R. Roesser and distributed by the General Electric Company of America. This presentation is sometimes referred to as a Segrè plot

N

1.5 Constitution of the nucleus

We have seen in section 1.3 that the atom of an isotope of mass number A is quite closely A times the mass of the lightest hydrogen isotope. In so far as the mass of the atomic electrons is negligible compared to the mass of the nucleus, this means that the mass of the nucleus is about A times the mass of the nucleus of the lightest hydrogen isotope. The latter nucleus, which we would expect to be the simplest of all nuclei, is given the name *proton*; it has $A = 1$ and $Z = 1$. A simple hypothesis which suggests itself is that the general nucleus consists of A protons and $A - Z$ electrons, thus having a mass approximately A times the mass of the proton, and a net positive charge of Ze. However, as we shall see when we come to discuss intrinsic angular momentum or 'spin', such a hypothetical assembly in certain cases does not have a spin in agreement with the measured nuclear spin. There are additional objections based on the high energy that electrons would necessarily have if confined to a volume of nuclear dimensions. With the discovery of the *neutron* (Chadwick, 1932) another and, as it proved, satisfactory hypothesis could be made. The neutron has no charge and has a mass almost, but not quite, equal to that of the proton. It is convenient to introduce *nucleon* as a generic term for a particle which is either a proton or a neutron. We can then consider a nucleus to be built from Z protons and N neutrons. It will have a mass number A provided $Z + N = A$. In other words it contains A nucleons. It is in these terms that the nucleus is presently pictured.

Any two of the three integers Z, N, A completely determine the nuclear species or *nuclide* under discussion. A nuclide is usually denoted by its chemical symbol with a prefix indicating the A-value. A second lower prefix may be used to indicate the Z-value, though this of course is not necessary because the Z-value is already fixed by the chemical symbol. Thus the lighter of the stable helium isotopes is denoted by ^{3}He or by $^3_2$He.

A useful plot in common use on which to present nuclear properties is shown in part in Figure 1. With N as abscissa and Z as ordinate, each nuclide has its own square on the chart. Since the chemical behaviour is determined by the number of orbiting electrons (i.e. the Z-value) all the isotopes of a given chemical element, having the same Z-value, lie on a horizontal line. Nuclides having the same N-value, called *isotones*, lie on a vertical line, while nuclides having the same A-value, called *isobars*, lie on a line with a backward slope of $45°$.

1.6 Units

Having outlined the nature, size and mass of the structure, the atomic nucleus, which is the subject of our study, we conclude these introductory paragraphs with a few remarks concerning the units to be employed.

The praiseworthy attempt to achieve a standardized set of units in all scientific applications by the adoption of the SI (*Système International d'Unités*) units (International Conference on Weights and Measures 1960) has not yet led to the displacement of the units introduced by the pioneers in nuclear physics. The natural resistance to change, however trivial, in mental habits is supported in some instances by the requirement for special units (for example nuclear magneton) which are not likely to have applications in outside fields. The relationships to the SI units are usually straightforward and these are included in Appendix D.

It is a happy coincidence that the unit of *length*, the *femtometer* (10^{-15} m), abbreviated to fm, which is the order of size of the nucleus, allows, without causing confusion, the continuation in the use of the name 'fermi' for this unit in honour of the Italian physicist who pioneered major developments in both theoretical and experimental nuclear physics.

The continued development of experimental techniques has brought the requirement for the general use of decreasingly smaller units of time. Having left microseconds (10^{-6} s) and nanoseconds (10^{-9} s) behind we are passing into, and no doubt through, the era of picoseconds (10^{-12} s). As will later be discussed, even this unit is long on the natural nuclear time scale.

The size of the useful *energy* unit is set by the binding energy of the nuclear components. This is about a million times greater than the binding energy of the valence atomic electrons. The unit which had been used in atomic physics was the electronvolt, defined as the change in kinetic energy of an electron moving in vacuum through a potential difference of one volt. It was therefore natural in nuclear physics to adopt a million electronvolts (MeV) equal to 10^6 eV as the unit of energy.

The *charge* on the nucleus is customarily expressed in electrostatic units (e.s.u.). This leads to the necessary introduction of the velocity of light in some symbolic electromagnetic formulae (e.g. nuclear magneton) which have become traditional in the subject. The only other unit we need mention is the gauss (G), which is used for the quantity loosely referred to as *the magnetic field*. This expression will normally refer to the magnetic flux density, the SI unit for which is the weber per metre squared which equals 10^4 gauss. The reader is referred to *Basic Electricity* by W. M. Gibson in the Penguin Library of Physical Sciences for a discussion of these and the other electromagnetic units.

Chapter 2
Radioactive Decay Laws

2.1   Introduction

The fact that the atoms of most materials do not spontaneously change their nuclear or chemical properties indicates that certain assemblies of nucleons with particular Z- and N-values, under the action we must assume of internal attractive forces, form stable structures. By no means all nucleon assemblies are stable. Some are so far from stability that, immediately after formation, one or more nucleons may be emitted and the grouping changed. However there exist certain assemblies that are very close to stability but are still not strictly speaking stable. They do not have enough excess energy to permit the immediate emission of even one nucleon and they exist for appreciable times before stabilizing by undergoing a transformation constituting one of the modes of radioactive decay and thereby altering their Z- and N-numbers. Before discussing in detail the various modes of radioactive decay, from which we derive considerable insight into the physical conditions within and around the nucleus, it is convenient to collect together certain mathematical results which are applicable to all modes of radioactive decay.

2.2   The exponential decay law

We start from the assertion that the probability that a particular nucleus undergoes radioactive decay in a time interval dt is λdt, where, for a particular nuclide, λ is a constant called the *decay constant*. This assertion implies that the previous history of the nucleus has no bearing on its probability of decay at a particular time. For example, it means that when radioactive nuclei are being produced artificially, in, say, a reactor over a period of time, the 'older' nuclei at any instant are no more likely to decay than those more recently formed.

If now we consider a large number N of similar radioactive nuclei in a sample and if $-dN$ of these decay in an interval dt then, providing N is very large, by definition the probability that any one nucleus decays in the interval is $-dN/N$. From the definition of the decay constant this probability is also $\lambda \, dt$. Therefore

$$\frac{dN}{dt} = -\lambda N.$$

The general solution of this equation is

$\ln N = -\lambda t + \text{constant}.$

Choosing the constant of integration to suit the initial condition that there be N_0 nuclei originally in the sample, we have

$$N = N_0 e^{-\lambda t}. \tag{2.1}$$

Experimentally, radioactivity is usually investigated by detecting the products of the decay. It is thus the number of decays per unit time, defined as the *activity* of the sample, which is of experimental relevance. In the present case, if we denote the activity by A, then by definition

$$A = -\frac{dN}{dt} = \lambda N_0 e^{-\lambda t},$$

from **2.1**.

This may be written

$$A = A_0 e^{-\lambda t}, \tag{2.2}$$

where A_0 is the initial activity.

The justification for the original assertion will now lie in the accuracy with which this exponential law describes the observed decrease in activity with time. In practice it may be tested by plotting $\ln A$ against t, when, if the assertion is valid, a straight line of negative slope λ should result. In the case of sources of reasonably high activity, such a linear plot is found to result and this is a recognized way of determining λ.

2.3   Half-life and mean life

Very often, instead of λ, an equivalent quantity, *the half-life*, denoted by $T_{\frac{1}{2}}$ is used. It is defined as the time taken for the activity to fall to one half of its original value. From this definition and equation 2.2 we see that

$$\tfrac{1}{2}A_0 = A_0 e^{-\lambda T_{\frac{1}{2}}},$$

and therefore $$T_{\frac{1}{2}} = \frac{\ln 2}{\lambda} = \frac{0\cdot693}{\lambda}. \tag{2.3}$$

$T_{\frac{1}{2}}$ has the dimension of time; λ has the dimension of reciprocal time.

The *mean life* τ, in the ordinary sense of the average life of all nuclei in a large assembly, is also of interest. The number of nuclei in the assembly having life t will be the number which decay in the interval from t to $t + dt$. This is $\lambda N(t) \, dt$. Forming the average in the usual way we then have

$$\tau = \frac{\int_0^\infty \lambda N(t) t \, dt}{N_0}.$$

Substituting for $N(t)$ from equation 2.1 and carrying out the integration it follows that

$$\lambda\tau = 1$$

and $\quad \tau = \dfrac{T_{\frac{1}{2}}}{\ln 2}$.

2.4 Series decay

It may be that the *daughter* nucleus, which is formed as the result of the decay of the *parent* nucleus, is itself radioactive with its own decay constant and half-life. Its subsequent decay may be experimentally distinguishable from that of its parent. In this circumstance, the two activities can be measured separately as functions of time.

Let $N(t)$ be the number of parent nuclei present at a given time and $n(t)$ the number of daughter nuclei at the same time. Let λ_P and λ_D be the respective decay constants. Then, as before, for the parent

$$dN(t) = -\lambda_P N(t) \, dt. \tag{2.5}$$

For the daughter, however, the population is now increased by the decay of the parent as well as decreased by its own decay. Thus

$$dn(t) = -\lambda_D n(t) \, dt + \lambda_P N(t) \, dt. \tag{2.6}$$

From equation 2.5, as before,

$$N(t) = N_0 e^{-\lambda_P t}.$$

Substituting this value into equation 2.6 and multiplying throughout by $\exp(\lambda_D t)$ we have

$$\frac{d}{dt}[e^{\lambda_D t} n(t)] = \lambda_P N_0 e^{(\lambda_D - \lambda_P)t}.$$

It follows by integration that

$$e^{\lambda_D t} n(t) = \frac{\lambda_P N_0}{\lambda_D - \lambda_P} e^{(\lambda_D - \lambda_P)t} + \text{constant}.$$

If now the original conditions are that no daughter nuclei are present, i.e. $n(0) = 0$, then the constant of integration is such that

$$n(t) = \lambda_P N_0 \left[\frac{e^{-\lambda_P t}}{\lambda_D - \lambda_P} + \frac{e^{-\lambda_D t}}{\lambda_P - \lambda_D} \right]. \tag{2.7}$$

Since $n(t)$ starts from zero at $t = 0$ and must fall to zero as t approaches infinity, there must be at least one intermediate maximum value. Differentiating equation 2.7 with respect to t and equating dn/dt to zero we find

$$t_{max} = \frac{\ln(\lambda_P/\lambda_D)}{\lambda_P - \lambda_D}. \tag{2.8}$$

It is important to note that in this case the activity is *not* equal to $dn(t)/dt$. The number of nuclei *decaying* in time dt is not $dn(t)$ as given in equation 2.6 but only the first term on the right-hand side of that equation. Thus

$$a(t) = \lambda_D n(t) = \lambda_D A_0 \left[\frac{e^{-\lambda_P t}}{\lambda_D - \lambda_P} + \frac{e^{-\lambda_D t}}{\lambda_P - \lambda_D} \right]. \qquad 2.9$$

The activity of the daughter, therefore, also builds up from zero to reach a maximum value at a time given by equation 2.8, then falls off as the difference between two exponentials.

Two cases of practical interest arise.

2.4.1 Short-lived parent

If the half-life of the parent is very short in comparison with that of the daughter nuclide, then $\lambda_P \gg \lambda_D$. In this case $\exp(-\lambda_P t)$ becomes quickly negligible compared to $\exp(-\lambda_D t)$. If we also neglect λ_D in the denominator of equation 2.9 then the equation giving the long-term behaviour of the daughter activity is

$$a(t) = \frac{\lambda_D}{\lambda_P} A_0 e^{-\lambda_D t}. \qquad 2.10$$

Note that to this approximation the parent nuclei transform instantaneously into the daughter nuclei.

In Figure 2 the parent and daughter activities corresponding to the case where $\lambda_P = 10 \times \lambda_D$ are plotted as functions of time.

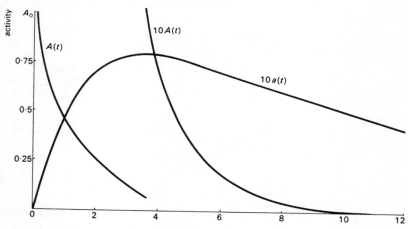

Figure 2 Plot of the variation with time of the radioactivities of short-lived parent and long-lived daughter materials in a source which initially contained only the parent material. The half-lives are taken to be in the ratio 1:10

2.4.2 Long-lived parent

If the parent has a much longer half-life than the daughter, then $\lambda_P \ll \lambda_D$. In this case, as t increases from zero, $\exp(-\lambda_D t)$ becomes quickly negligible compared with $\exp(-\lambda_P t)$. Also, we can neglect λ_P in the denominator of the right-hand side of equation 2.7 and write

$$n(t) = \frac{\lambda_P}{\lambda_D} N_0 e^{-\lambda_P t},$$

as the expression for $n(t)$ for large values of t. The activity of the daughter material is given as above by $\lambda_D n(t)$, i.e.

$$a(t) = \lambda_P N_0 e^{-\lambda_P t}.$$

This expression is of course also equal to A_t. Thus to the extent that the approximations are valid, the activities of daughter and parent materials are equal, and the rate of creation and rate of decay of the daughter material balance. It is not however an equilibrium condition as these rates are slowly changing with time. The term *secular equilibrium* is used to describe the situation in this radioactive context. When secular equilibrium has been attained, note that the ratio of the amounts of daughter and parent materials present is the ratio of their half-lives.

In Figure 3 the activities of parent and daughter, in the case when $\lambda_P = \frac{1}{10}\lambda_D$, are plotted as functions of time.

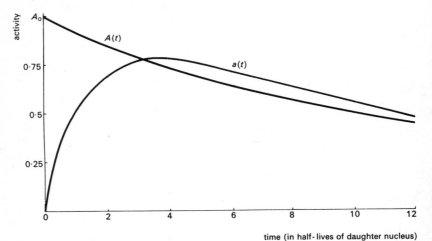

time (in half-lives of daughter nucleus)

Figure 3 Plot of the variation with time of the radioactivities of long-lived parent and short-lived daughter materials in a source which initially contained only the parent material. The half-lives are taken to be in the ratio 10:1

2.5 Naturally occurring radioactive series

The theory developed in section 2.4 can be applied to the naturally occurring radioactive series. In the case of each of these series a very long-lived isotope (^{232}Th, half-life $1\cdot41 \times 10^{10}$ years in the case of the thorium series, ^{238}U, half-life $4\cdot51 \times 10^{9}$ years in the case of the uranium series and ^{235}U, half-life $7\cdot07 \times 10^{8}$ years in the case of the actinium series) constitutes the parent. From this parent stem between ten and twenty generations of radioactive descendants in each case. Above we have analysed the case involving only one descendant. However an equation similar to equation 2.6 can be formed for each succeeding generation. The mathematical analysis may then be carried out exactly as above and, on the basis that all half-lives are very much shorter than that of the parent material, secular equilibrium will be established. When sufficient time has elapsed for this equilibrium to be attained, all the members of the series have equal activities and the amount of material associated with any member is proportional to its half-life.

2.6 Definition of the curie

As a further application of these ideas, we consider the definition of the unit of activity named the *curie*. Originally the curie was defined as the activity of that amount of ^{222}Rn (radon; half-life $3\cdot825$ days) which is in secular equilibrium with one gramme of ^{226}Ra (radium; half-life 1622 years). The unit was so defined to permit sources of standard activity to be produced wherever a radium sample of known weight was available. It follows, since the radon is in secular equilibrium, that its activity will be the same as that of the gramme of radium. The curie is thus equal to

$$\lambda N_0 = \frac{0\cdot693}{T_{\frac{1}{2}}} \frac{N_A}{A},$$

where N_A is Avogadro's constant, A the atomic weight of radium and $T_{\frac{1}{2}}$ the half-life of radium. On substituting the numerical values for these quantities the curie may be seen to be $3\cdot61 \times 10^{10}$ disintegrations per second. However, to make the unit independent of the half-life of radium, redetermination of which had several times necessitated changing the curie as a practical unit, an internationally agreed definition of the curie as $3\cdot7 \times 10^{10}$ disintegrations per second is now accepted and the traditional unit abandoned.

2.7 Branching or parallel decay

It may happen that a nucleus can decay by either of two modes. If the probability per unit time that it will decay by mode one is λ_1 and by mode two is λ_2, then the probability that it decays by one *or* by two is $\lambda_1 + \lambda_2$. This latter quantity will be the decay constant on the usual definition. λ_1 and λ_2 are termed the *partial decay constants* and $\lambda = \lambda_1 + \lambda_2$ the *total decay constant*. These terms could in the obvious way be extended to more than two competing decay modes if necessary.

27 Branching or parallel decay

It is possible that the activity corresponding to decay by mode one can be measured without interference from the activity arising from mode two. For example, ^{64}Cu decays by one mode to ^{64}Ni and by another to ^{64}Zn. A detecting system can be set up to detect only the decays to ^{64}Ni. In this case, the activity measured $A_1(t)$ will be given by $\lambda_1 N(t)$. Now

$$dN(t) = -\lambda_1 N(t)\, dt - \lambda_2 N(t)\, dt = -(\lambda_1 + \lambda_2) N(t)\, dt = -\lambda N(t)\, dt,$$

from which it follows that $N(t) = N_0 e^{-\lambda t}$ and

$$A_1(t) = \lambda_1 N(t) = \lambda_1 N_0 e^{-\lambda t} = (A_1)_0 e^{-\lambda t}.$$

The half-life exhibited is therefore that corresponding to the total (not the partial) decay constant, and the material, in this case ^{64}Cu (despite its different possible decay modes), is still characterized by only one half-life. If it is desired to measure λ_1 or λ_2, then the fraction of the total number of decays proceeding by the mode in question must be determined.

2.8 Artificial radioactivity

It is possible to induce radioactivity in an initially nonradioactive sample by subjecting it to neutron irradiation in a reactor, or to particle bombardment in an accelerator. Let it be assumed that the production of the radioactive nuclei by one or other of these methods proceeds at a constant rate of S nuclei per unit time. If $N(t)$ is the total number of radioactive nuclei at time t, then

$$dN(t) = S\, dt - \lambda N(t)\, dt,$$

the right-hand side expressing the competition between increase due to production and decrease due to radioactive decay. This relation leads immediately to the differential equation

$$\frac{dN(t)}{dt} = S - \lambda N(t),$$

which on integration yields

$$N(t) = \frac{S}{\lambda} + \text{constant} \times e^{-\lambda t}.$$

If the starting conditions are that there is no activity in the sample, then $N(0) = 0$ and the constant of integration is such that

$$N(t) = \frac{S}{\lambda} [1 - e^{-\lambda t}].$$

The activity at any time is given by

$$A(t) = \lambda N(t) = S[1 - e^{-\lambda t}]. \qquad\qquad 2.11$$

As t gets very large, clearly the situation is reached of a balance between the activity, that is the decay rate, and S, the production rate. This means that the activity produced can never exceed S and approaches it asymptotically. Note that one half-life from the beginning of the irradiation the activity reaches one half of its asymptotic value, after two half-lives it reaches three quarters. It is seldom economic to prolong irradiations beyond this stage in view of the diminishing increases in activity.

It is usual, instead of specifying S, to define the *yield Y* in an irradiation of this kind. The yield is the rate of increase of activity at time $t = 0$, i.e.

$$Y = \left[\frac{dA}{dt} \right]_{t=0}.$$

The activity-against-time curve, shown in Figure 4, thus starts off from a zero value at $t = 0$ with a tangent of slope Y. The value of Y is, from differentiating equation **2.11**, given by $Y = \lambda S$ and equation **2.11** may then be written

$$A(t) = \frac{Y}{\lambda} [1 - e^{-\lambda t}].$$

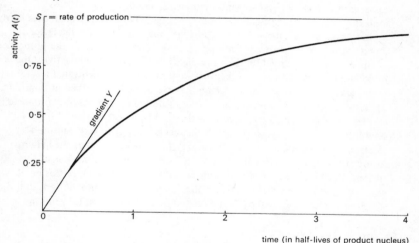

Figure 4 Artificially produced radioactivity as a function of time of bombardment; S is the rate of production, Y the yield

2.9   Summary

The assumption that a radioactive nucleus has a transformation probability per unit time which is independent of its previous history enabled a formula to be derived which related the activity of a simple source to the time of observation. This dependence, the 'exponential law', was found to be in strict agreement with experiment. Equally valid formulae were derived for the cases of series and parallel decay arising in the naturally occurring radioactive materials and for the case of the production of radioactivity in accelerators and nuclear reactors.

Chapter 3
Radioactivity: Alpha Decay

3.1 Fundamentals of alpha decay

Certain radioactive nuclei, on transforming, emit positively charged particles whose measured charge and e/m values indicate that their charge is twice, and their mass four times, that of the proton. These particles, on passing through a gas or into a solid material, expend their energy rapidly in the process of ionization (i.e. stripping electrons from the originally neutral atoms), thus leaving in their wake a short, dense track of positive ions and electrons, the latter remaining as free electrons or forming negative ions depending on the properties of the medium concerned. In the early days of the study of radioactivity these heavy doubly charged particles, because they produced concentrated ionization in the gas of electroscopes and because they were readily absorbed in thin foils, again by virtue of their ionizing properties, were the first of the 'radiations' to be studied and were named *α-particles*. Rutherford's pioneer work of 1909 established that α-particles, after they had been brought to rest, captured electrons and became atoms of helium gas. It is now known that the helium isotope concerned is ^{4}He and hence the α-particle, being the nucleus of this isotope, must be a cluster of two protons and two neutrons. Once this is appreciated it follows that in α-decay the parent nucleus, denoted by P, and the daughter nucleus, denoted by D, are related in A, Z and N value according to

$$^A_Z P_N \rightarrow {}^{A-4}_{Z-2} D_{N-2} + {}^4_2 He_2.$$

This relationship, which was in the early days of radioactivity referred to as the *displacement law*, may also be stated as $\Delta A = -4, \Delta Z = -2, \Delta N = -2$.

3.2 Mass-energy relations in alpha decay

The conservation of mass–energy must apply to the α-decay process. Hence

$$M'_P = M'_D + M'_\alpha + Q,$$

where the masses are the masses of the nuclei and Q is the energy shared by the products of the reaction. As in all equations of this type with which we shall be concerned, we can either write the quantities in the traditional mass units or write all quantities in energy units, mass and energy transforming according to $E = mc^2$.

For the process to be energetically possible Q must be positive and it therefore follows that

$$M_P' > M_D' + M_\alpha'.$$

Universally, nowadays, in the published mass tables, neutral-atom masses rather than nuclear masses are listed. The inequality is therefore customarily written as $M_P > M_D + M_{He}$ where the masses are now neutral-atom masses. In taking this step it is however to be noted that something more than adding Z electrons to each side of the inequality is involved. $M_P < M_P' + Zm_e$ by an amount equal to that mass which is equivalent to the binding energy of the Z electrons in the atom. For heavy nuclei such as those involved in α-decay, this total binding energy is of the order of several hundred thousand electronvolts. However, this is largely compensated by the total binding energy of the atoms involved on the other side of the inequality and the overall error introduced in substituting atomic for nuclear masses is usually negligible compared with the α-particle energies, which are in the range of a few million electronvolts.

From tables of atomic masses, it is found that for A-values from about 150 upwards very many nuclei should be unstable against α-decay. For reasons which are discussed in section 3.16 the mass condition being satisfied does not necessarily result in observable α-activity. The plot in Figure 5 of the distribution of α-emitters as a function of A-values indicates the extent to which the α mode of decay is limited to the heavy nuclides.

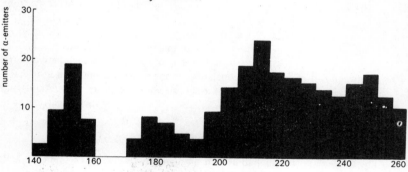

Figure 5 Histogram of the number of α-emitters in terms of their A-values (based on *Chart of the Nuclides,* 1966)

3.3   **Alpha-particle fine structure**

The kinetic energy of the α-particles emitted by a particular isotope can be measured accurately with a magnetic spectrometer. When such measurements are made, it is found that in some cases there is only one group of monoenergetic -particles; in other cases there are two or more such groups. When there is more than one group, the α-particle spectrum is said to exhibit *fine structure*, a phenomenon which is discussed in detail in section 3.15.

31   Alpha-particle fine structure

3.4 Recoil energy in alpha decay

The uniqueness of the α-particle energy in the general case points to the process being a 'two-body' process with the available energy Q shared by the α-particle and the daughter nucleus, which must recoil to conserve linear momentum. As the kinetic energies involved are no more than a few million electronvolts, whereas the rest masses are to be measured in thousands of millions of electronvolts, we may without appreciable error assume the formulae of Newtonian dynamics to apply. Thus, if M_D and M_α are the masses of the daughter nucleus and α-particle and T_D, T_α, V_D, V_α their kinetic energies and velocities, $M_D V_D = M_\alpha V_\alpha$ by the conservation of linear momentum. Therefore

$$T_D = \tfrac{1}{2} M_D V_D^2 = \frac{M_\alpha}{M_D} T_\alpha.$$

Now
$$Q = T_D + T_\alpha = T_\alpha \left[1 + \frac{M_\alpha}{M_D} \right],$$

so we have
$$T_\alpha = \frac{M_D}{M_D + M_\alpha} Q \simeq \frac{M_D}{M_P} Q \quad \text{and} \quad T_D \simeq \frac{M_\alpha}{M_P} Q. \qquad 3.1$$

3.5 Observed energies and half-lives of alpha emitters

The energies of α-particles emitted by different α-active nuclides range from 1·9 MeV for ^{144}Nd to 9·2 MeV for ^{213}At. The half-lives range from 2×10^{17} yr in the case of ^{209}Bi to $2·9 \times 10^{-7}$ s in the case of ^{212}Po. Thus, whereas the energies are contained within a range of one order of magnitude, the half-lives range over more than thirty orders of magnitude. There is an apparent correlation between half-life (or decay constant) and the α-particle energy; a short half-life is associated with a high value of α-particle energy and vice versa. An attempt to fit the then known (1912) experimental α-decay constants and particle energies by an empirical formula of the form

$$\log \lambda = C_1 \log T_\alpha + C_2, \qquad 3.2$$

by Geiger and Nuttall, had limited success. While values of C_1 and C_2 could be found to give a satisfactory fit of the calculated decay constant with the measured decay constant within one natural radioactive series, the values of the constants had to be altered to maintain the goodness of the fit on going from one radioactive series to another. Insight into the correlation between half-life and α-particle energy had to await the development of the theory of α-decay, which we now proceed to discuss.

3.6 Rutherford scattering: theoretical treatment

To be in a position to consider a 'model' in terms of which to picture α-decay we require quantitative information about the electrostatic field around the nucleus. This can be obtained from the study of scattering in this field of incoming α-particles.

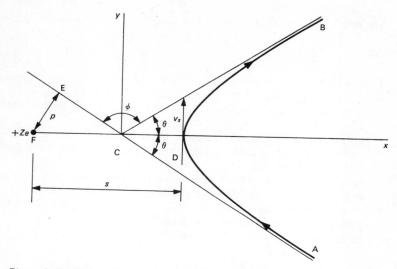

Figure 6 Scattering of an α-particle by a nucleus, charge number Z, with an impact parameter p

We begin by considering the motion of a single α-particle in the Coulomb field of a single nucleus which is assumed to be fixed in position. In Figure 6 the incoming α-particle is directed along AC. Under the influence of the Coulomb field associated with the charge Ze on the nucleus situated at the point F, the particle will experience a repulsive force whose line of action passes through F and whose magnitude is inversely proportional to the square of the distance of the particle from F. This is completely analogous to planetary motion under gravity.† The trajectory is a conic section, in this case a branch of the hyperbola with the nucleus at a focus F. The α-particle will be scattered through an angle ϕ and goes off finally along CB.

Let A and B be the lengths of the principal semi-axes of the hyperbola. Referred to the principal axes Cx and Cy, the trajectory has the equation

$$\frac{x^2}{A^2} - \frac{y^2}{B^2} = 1,$$

and the asymptotes AC and BC have the equations

$$\frac{x}{A} \pm \frac{y}{B} = 0.$$

From the gradient of these lines, it follows that

$$\tan \theta = \frac{B}{A}. \tag{3.3}$$

† For gravity, as for charged particles of opposite sign, the force is attractive, and the other branch of the hyperbola is followed.

The eccentricity of the hyperbola, ϵ, is given by

$$\epsilon^2 = \frac{A^2 + B^2}{A^2}.$$ 3.4

From the geometry of the hyperbola CF = ϵA and CD = A. Thus s, the distance of closest approach to the nucleus for the trajectory under consideration is given by s = FD = $A(1 + \epsilon)$. From equations 3.3 and 3.4 we see that $\epsilon = \sec \theta$ and hence $s = A(1 + \sec \theta)$.

It is customary to specify the 'closeness' of the collision by the *impact parameter* which is p, the length of the perpendicular from the scattering centre on to the original direction of travel of the scattered particle. A 'head-on' collision corresponds to $p = 0$, the collision getting more 'distant' as p increases. We wish to derive a relation between the angle of scatter ϕ and the impact parameter p. We start by equating the total energy (i.e. kinetic plus potential) at the point D on the trajectory with the kinetic energy at infinity, where the potential energy is zero by definition. Thus

$$\tfrac{1}{2}M_\alpha V_s^2 + \frac{2Ze^2}{s} = \tfrac{1}{2}M_\alpha V_0^2.$$ 3.5

The line of action of the force passing through F, the angular momentum of the particle about an axis through F must stay constant throughout the motion. Thus

$$M_\alpha V_0 p = M_\alpha V_s s.$$

From this it follows that

$$V_s = \frac{p}{s} V_0.$$

On substituting this value of V_s into equation 3.5 we have

$$\tfrac{1}{2}M_\alpha V_0^2 \left(\frac{s^2 - p^2}{s} \right) = 2Ze^2.$$

It is convenient to introduce b, the distance of closest approach in a head-on collision. By equating the potential energy at a separation b, at which the particle is instantaneously at rest, to the kinetic energy at infinity, we have

$$b = \frac{4Ze^2}{M_\alpha V_0^2}.$$

Hence $\quad p^2 = s(s - b).$ 3.6

Now we wish to find a relation between s and θ. From the triangle FCE, we see that $\sin \theta = p/A\epsilon$. From the values of ϵ and A given above it follows that

$$\epsilon = \sec \theta \quad \text{and} \quad A = \frac{s}{1 + \sec \theta}.$$

Hence $\quad \sin\theta = \dfrac{p(1 + \sec\theta)}{s\sec\theta}$

and $\quad s = p\cot\tfrac{1}{2}\theta$.

3.7

When this is substituted in equation **3.6** we find

$p = \tfrac{1}{2}b\tan\theta$.

Since $\phi = \pi - 2\theta$ we have finally

$p = \tfrac{1}{2}b\cot\tfrac{1}{2}\phi$.

3.8

We now pause in the discussion to examine the values of s and b arising in the experimental situation and to establish their magnitudes relative to atomic dimensions. Suppose an α-particle of energy 4·2 MeV, emitted by a $^{238}_{92}U$ nucleus to be scattered by another $^{238}_{92}U$ nucleus. In this case

$$b = \frac{2 \times 92 \times (4\cdot8 \times 10^{-10})^2}{4\cdot2 \times 1\cdot6 \times 10^{-6}} \ \text{m} = 63 \ \text{fm},$$

the fermi, defined as 10^{-15} m, being a suitable unit of length in a nuclear context. Using this value in equation **3.8** we see that for scattering angles in excess of $20°$, the impact parameter must be less than 180 fm. Hence, from equation **3.7**, s must be less than 215 fm. The radius of the orbit of the two innermost electrons (i.e. the K-electrons) in the uranium atom is about 600 fm. Thus we see that the α-particle in the energy range being used experiences the full effect of the nuclear charge during the important part of the collision.

We now resume the general discussion of scattering by considering the experimental arrangement sketched in Figure 7. A well-collimated beam of α-particles falls at right angles on a thin foil, thickness t, containing n scattering centres (say nuclei with charge $+Ze$) per unit volume. The foil is assumed to be so thin that the loss of α-particle energy by ionization is negligible. In these

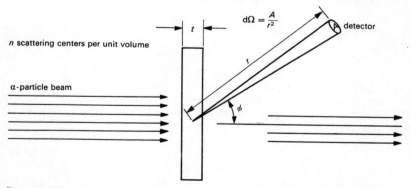

Figure 7 The geometry of the scattering of an α-particle beam by a thin foil

circumstances, the probability of an α-particle experiencing two large-angle-scattering collisions is also negligible. The spot on which the α-particles are incident is viewed at an angle ϕ to the beam by an α-particle detector subtending a solid angle of $d\Omega$ for scattered particles. We now consider how the rate of detection of α-particles will be expected to vary with the angle ϕ.

The probability that an α-particle be scattered through an angle lying between ϕ and $\phi + d\phi$ will equal the probability of finding a scattering centre at a distance between p and $p + dp$ from the α-particle trajectory. This probability is to be measured by the average number of nuclei (necessarily very much less than unity) contained in the volume of a hollow cylinder of radius p, thickness dp and length t. This volume is equal to $2\pi pt\, dp$ and hence the required probability is $2\pi pnt\, dp$. If A is the number of incident α-particles per unit time, the number scattered through an angle between ϕ and $\phi + d\phi$ per unit time is $2\pi Apnt\, dp$. These particles will be scattered into a hollow cone having an inside semi-angle equal to ϕ and an outside semi-angle equal to $\phi + d\phi$ with its axis along the direction of the incident α-particles. The scattered particles are thus contained in a total solid angle $2\pi \sin\phi\, d\phi$. The fraction of the scattered particles which enter the detector is then the ratio of the solid angles, namely

$$\frac{d\Omega}{2\pi \sin\phi\, d\phi}.$$

Therefore the number of particles detected per unit time is equal to

$$C = \frac{2\pi Apnt\, dp\, d\Omega}{2\pi \sin\phi\, d\phi}. \qquad 3.9$$

But, from equation 3.8,

$$\left|\frac{dp}{d\phi}\right| = \tfrac{1}{4}b \cosec^2 \tfrac{1}{2}\phi.$$

Substituting this value in equation 3.9 we have

$$C = A\frac{b^2}{16}\, nt \cosec^4 \tfrac{1}{2}\phi\, d\Omega. \qquad 3.10$$

If now it is found in a scattering experiment with α-particles of a given energy that the dependence of C on ϕ is accurately described by equation 3.10 for angles greater than a few degrees, then we can conclude that the potential down to a separation distance of b is accurately proportional to $1/r$, i.e. it is the Coulomb potential. Any other dependence of potential on r would necessarily lead to a different angular distribution.

We now note that the derivation of equation 3.10, above, rests on the total charge of the nucleus Ze being effective, and therefore the impact parameter must not exceed 600 fm, the radius of the K-shell. This restriction on p-value means that a restriction on the ϕ-value follows from equation 3.8. It means that in the case of uranium ϕ must be greater than 6°. It is also to be noted that the foil thickness permitted, having regard to the requirement to limit the scattering to 'pure single scattering', is given by the condition that $2\pi pnt\, dp$ is very much less than unity. This means that t must be very much less than

$$\frac{1}{2\pi pn\, dp} = \frac{4}{2\pi pnb\, \mathrm{cosec}^2\, \tfrac{1}{2}\phi\, d\phi}.$$

Assuming there to be about 5×10^{19} scattering centres per millimetre cubed and that the detector accepts an angular range of scattered particles of about a tenth of a radian, then the single scattering condition will be satisfied up to a foil thickness of 10^{-2} mm.

It should be noted that we have neglected in the above derivation the effect of the recoil of the scattering nucleus. This is equivalent to assuming that the nuclear mass is infinite. A more general treatment by C. Darwin (1914) shows that formula 3.10 holds for finite nuclear mass provided:

(a) that the α-particle mass used to evaluate b is replaced by the expression

$$\frac{m_\alpha\, m_{\mathrm{nucleus}}}{m_\alpha + m_{\mathrm{nucleus}}},$$

which is referred to as the *reduced mass*, and

(b) that the angle ϕ, the scattering angle measured in the laboratory reference frame, is replaced by the scattering angle measured in a frame of reference travelling with the centre-of-mass of the α-particle and nucleus. This frame of reference will have a constant velocity

$$\frac{m_\alpha}{m_\alpha + m_{\mathrm{nucleus}}}\, V_\alpha,$$

with respect to the frame of reference fixed in the laboratory, and we refer to the moving frame as the *centre-of-mass system*.

3.7   **Rutherford scattering: experimental results**

Geiger and Marsden (1913), by scattering α-particles of 7·68 MeV in gold films typically 3×10^{-4} mm thick, confirmed formula 3.10 over a range of values of ϕ from $5°$ to $150°$. The distance of closest approach for an α-particle of this energy to a gold nucleus, in the event of a scatter of $150°$, is, from equations 3.8 and 3.7, 30 fm. It was thus established that from the dimensions of the radius of the K-shell down to 30 fm the law of force is accurately that for the Coulomb field surrounding a point charge. It was on the basis of this observation and by the reasoning given above that Rutherford proposed the atomic model that now bears his name. The gold foil of the thickness used by Geiger and Marsden is seen, in the light of the above discussion, to be such that the probability of double scattering can be neglected. Also the energy loss of the α-particle by ionization in passing through a foil of this thickness is a negligibly small fraction of its kinetic energy, and hence the α-particle energy may be assumed constant throughout its passage through the foil.

Later experiments with uranium foils gave angular distributions in agreement with the Rutherford formula and showed that in the case of uranium the Coulomb law held down to at least a separation of thirty fermis.

3.8 Alpha-decay paradox

The α-particle-scattering experiments showed that the field around the nucleus, $Z = 90$, was accurately the Coulomb field of a point charge down to thirty fermis from the centre of the nucleus. Any α-particle emitted by radioactive decay must therefore originate from a point closer to the nuclear centre than thirty fermis. Hence it must emerge with at least the electrostatic potential energy a doubly charged particle would have at that distance from the nucleus. This is about 8·6 MeV. However, α-particles of about 4 MeV are observed to be emitted from nuclei with $Z = 92$. This paradox could not be resolved within the framework of classical physics; understanding had to await the proposal of Gamow (and independently of Condon and Gurney) in 1928, who suggested abandoning the classical description and substituting one in terms of wave mechanics.

3.9 Nuclear potential barrier

Let us consider the potential energy of an α-particle as a function of its separation, measured centre-to-centre, from a heavy nucleus. Scattering experiments of the type discussed above have now been extended to higher energies and have established that, up to α-particle energies of 28 MeV, formula **3.10** holds. Thus for separations in excess of ten fermis the potential is that of the Coulomb force falling off as $1/r$. At about a distance of ten fermis the Rutherford scattering formula breaks down. This is due to forces between nucleons in the nucleus and the nucleons in the passing α-particle coming into play. These forces, which are believed to dominate inside the nucleus and which hold the nucleons together, the so called *strong-interaction* forces, are attractive and only operate for very short distances of separation. They cause a rapid fall in the potential when the α-particle gets within their short range. We make the simplifying assumption that the potential falls infinitely fast as the nuclear surface is crossed to reach a constant value inside the nucleus of $-V_0$, where $V_0 \ll B$ the maximum value the potential reaches. This potential curve is drawn in Figure 8. If we now think of this curve as the potential of the daughter nucleus, then the parent nucleus may be described by adding an α-particle which moves in the region $r \ll R$ with a kinetic energy $T_\alpha + V_0$ where T_α is the kinetic energy on emission (i.e. at infinite separation). The α-particle is represented in *total energy* (i.e. kinetic + potential) by the line ABCD on the diagram. On the basis of classical theory, the kinetic energy of the α-particle reaches zero at B and becomes negative between B and C. This is therefore a physically forbidden region which the α-particle cannot enter and hence an α-particle, once in the central well, is trapped forever in the central region. The potential is thus said to constitute a barrier which prevents the escape of the α-particle. In the wave-mechanical treatment however, the α-particle has a small but finite probability of penetrating into the region CB and in fact of succeeding in 'tunnelling' through the barrier to the region BA.

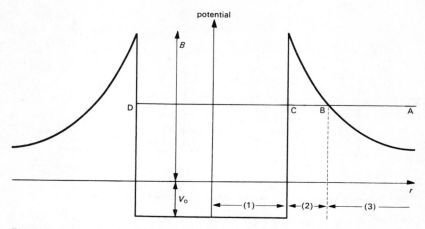

Figure 8 Potential well arising from short-range nuclear force combined with Coulomb potential due to long-range electrostatic force

Before treating a barrier in the shape of that in Figure 8 we consider a mathematically simpler but physically similar situation, namely a rectangular potential barrier in one dimension.

3.10 Rectangular potential barrier: one-dimensional wave-mechanical treatment

Consider a beam of particles incident on the barrier illustrated in Figure 9. In the wave-mechanical treatment (see, for example, R. M. Eisberg, *Fundamentals of Modern Physics*, Wiley, 1961, p. 212) the beam has an associated wave function Ψ, which is a function of x and t in the one-dimensional case. Ψ is the product of a time-dependent factor, which in the present problem is of the form $\exp(i2\pi\nu t)$ throughout, and a space dependent factor which we denote by $\psi(x)$. In any region of space $\psi(x)$ must satisfy the time-independent Schrödinger equation

$$\frac{d^2\psi}{dx^2} + \frac{2M(W-V)}{\hbar^2}\psi = 0,$$ 3.11a

where M is the mass of the particles in the beam, V is the potential energy in the region and $W = T + V$, where T is the kinetic energy in the same region.

We find it convenient to write Schrödinger's equation as

$$\frac{d^2\psi}{dx^2} + k^2\psi = 0,$$ 3.11b

with $k^2 = \dfrac{2M(W-V)}{\hbar^2}$.

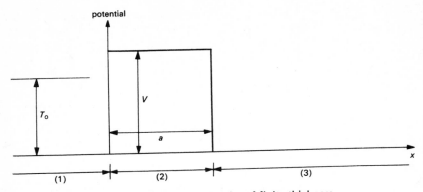

Figure 9 Idealized rectangular potential barrier of finite thickness

We now treat the three regions in Figure 9 in turn. In region (1) $V = 0$ and $W = T_0$. Therefore

$$k_1^2 = \frac{2MT_0}{\hbar^2} > 0.$$

The general solution of equation **3.11** then takes the form

$$\psi_1(x) = A_1 e^{ik_1 x} + B_1 e^{-ik_1 x}.$$

Taken in conjunction with the time-dependent factor $\exp(i2\pi\nu t)$, these terms then correspond to an incident wave travelling along the positive direction of the x-axis and a reflected wave travelling in the opposite direction.

In region (2) W is unchanged and therefore must still be taken as equal to T_0, and, as we are assuming that $T_0 < V$, it follows that $2M(T_0 - V)/\hbar^2$ is less than zero. Let it equal $-k_2^2$ where k_2 is real. Thus in region (2) equation **3.11b** becomes

$$\frac{d^2\psi}{dx^2} - k_2^2 \psi = 0,$$

which has the general solution

$$\psi_2(x) = A_2 e^{-k_2 x} + B_2 e^{k_2 x}.$$

Taken in conjunction with the factor $\exp(i2\pi\nu t)$ this is seen to correspond to standing waves.

In region (3) the situation with respect to W, V and T_0 is the same as in region (1). In this case however we need only consider the wave propagated along the positive direction of the x-axis as there is assumed to be no further potential discontinuities to cause reflections. Hence the solution in region (3) is

$$\psi_3(x) = A_3 e^{ik_1 x}.$$

The five constants A_1, B_1, A_2, B_2 and A_3 must now be chosen to achieve the correct conditions at the boundaries between the regions. We note that both

$W - V$ and the wave function, since it is related to the particle density in the beam, must everywhere have finite values. It follows from equation 3.11a that $d^2\psi/dx^2$ is everywhere finite. Hence $d\psi/dx$ cannot undergo sudden changes in value. In turn it can be argued that ψ cannot change discontinuously. Across the boundaries, ψ and $d\psi/dx$ must therefore both be continuous.

$\psi_1(0) = \psi_2(0)$ yields $A_1 + B_1 = A_2 + B_2$.

$$\left[\frac{d\psi_1}{dx}\right]_{x=0} = \left[\frac{d\psi_2}{dx}\right]_{x=0} \qquad \text{yields} \qquad ik_1\,A_1 - ik_1\,B_1 = -k_2\,A_2 + k_2\,B_2.$$

From $\psi_2(a) = \psi_3(a)$ we have $A_2\,e^{-k_2 a} + B_2\,e^{k_2 a} = A_3\,e^{ik_1 a}$, and from

$$\left[\frac{d\psi_2}{dx}\right]_{x=a} = \left[\frac{d\psi_3}{dx}\right]_{x=a}$$

we have $-k_2\,A_2\,e^{-k_2 a} + k_2\,B_2\,e^{k_2 a} = ik\,A_3\,e^{ik_1 a}$.

From these four equations B_1, A_2 and B_2 may be eliminated to give

$$\frac{A_1}{A_3} = \left[\frac{1}{2} + \frac{i}{4}\left(\frac{k_2}{k_1} - \frac{k_1}{k_2}\right)\right] e^{(ik_1 + k_2)a} + \left[\frac{1}{2} - \frac{i}{4}\left(\frac{k_2}{k_1} - \frac{k_1}{k_2}\right)\right] e^{(ik_1 - k_2)a}.$$

The flux of particles in the beam is given by the density of particles in the beam multiplied by the particle velocity. Since velocities in regions (1) and (3) are the same, the ratio of the fluxes in these regions will simply be in the ratio of the particle densities. This in turn, from wave-mechanical theory, is given by $|A_1/A_3|^2$. This may also be written $(A_1/A_3)^*(A_1/A_3)$, where $(A_1/A_3)^*$ is the complex conjugate of A_1/A_3. Taking the complex conjugate and carrying out the multiplication, we have

$$\left[\frac{A_1}{A_3}\right]^* \left[\frac{A_1}{A_3}\right] = \left[\frac{1}{4} + \frac{1}{16}\left(\frac{k_2}{k_1} - \frac{k_1}{k_2}\right)^2\right][e^{2k_2 a} + e^{-2k_2 a}] + \frac{1}{2} - \frac{1}{8}\left[\frac{k_2}{k_1} - \frac{k_1}{k_2}\right]^2.$$

Using the identity

$$\sinh^2 k_2\,a = \tfrac{1}{4}(e^{2k_2 a} + e^{-2k_2 a}) - \tfrac{1}{2},$$

and simplifying, we can write

$$\left[\frac{A_1}{A_3}\right]^* \left[\frac{A_1}{A_3}\right] = 1 + \frac{1}{4}\left[2 + \left(\frac{k_2}{k_1}\right)^2 + \left(\frac{k_1}{k_2}\right)^2\right]\sinh^2 k_2\,a. \qquad 3.12$$

But $\left[\dfrac{k_2}{k_1}\right]^2 = \dfrac{V - T_0}{T_0}.$

Substituting this in equation 3.12 we find

$$\left|\frac{A_1}{A_3}\right|^2 = 1 + \frac{1}{4} \frac{V^2}{T_0(V - T_0)} \sinh^2 k_2\, a.$$

The probability that an incident α-particle will penetrate the barrier is called the *penetration factor* and is equal to $|A_3/A_1|^2$. The ratio of the transmitted flux to the incident flux, which in the general case is given by

$$\left|\frac{A_3}{A_1}\right|^2 \frac{V_3}{V_1},$$

is called the *transmission coefficient*. As $V_3 = V_1$, these quantities are equal in the present case and we can write both equal to

$$\left|\frac{A_3}{A_1}\right|^2 = \left[1 + \frac{1}{4} \frac{V^2}{T_0(V - T_0)} \sinh^2 k_2\, a\right]^{-1}. \qquad 3.13$$

We recall that

$$k_2 = \frac{1}{\hbar} \sqrt{[2M(V - T_0)]}.$$

If $k_2\, a \gg 1$,

$$\sinh k_2\, a = \frac{e^{k_2 a} - e^{-k_2 a}}{2} \simeq \frac{e^{k_2 a}}{2},$$

and hence

$$\sinh^2 k_2\, a \simeq \frac{e^{2k_2 a}}{4}.$$

Also we note that

$$\frac{V^2}{4T_0(V - T_0)} \simeq 1$$

for $T_0 = \frac{1}{2}V$ and increases as T_0 decreases. We therefore proceed to ignore the first term in the bracket on the right-hand side of equation 3.13 and write

$$T = 16\, \frac{T_0}{V}\left[1 - \frac{T_0}{V}\right] e^{-2k_2 a},$$

where T is the transmission coefficient. The approximations made mean that this result is valid for a wide barrier which is high compared to the incident kinetic energy. Further, for the range of values of T_0/V normally of interest, which is T_0/V not too close to zero nor too close to unity, the factor

$$16\, \frac{T_0}{V}\left[1 - \frac{T_0}{V}\right]$$

lies between 1 and 4, and the value of T is dominated by the exponential factor. Therefore without serious error we can simply write

$$T = e^{-2\gamma},$$

where $\gamma = k_2 a$.

3.11 One-dimensional Coulomb barrier

We now consider the case of the one-dimensional potential illustrated in Figure 8. In region (2) we take the potential to be proportional to $1/x$ as for the Coulomb field. We assume that in this region the eigenfunction can be taken to be

$$\psi_2(x) = A_2 e^{-\gamma(x)} + B_2 e^{\gamma(x)}.$$

This is a generalization of the result for the rectangular barrier of constant height. In the present case the height is varying with distance through the barrier and the exponent cannot be assumed to be a linear function of x. We shall assume below that, since V is a slowly varying function of x, $\gamma(x)$ will also be slowly varying and $d^2\gamma/dx^2$ will consequently be very small. $\psi_2(x)$ must satisfy Schrödinger's equation, which in region (2), since $V = 2Ze^2/x$, will take the form

$$\frac{d^2\psi}{dx^2} + \frac{2M}{\hbar^2}\left[W - \frac{2Ze^2}{x}\right]\psi = 0. \qquad \qquad \textbf{3.14}$$

Now $\dfrac{d^2\psi_2}{dx^2} = \psi_2(x)\left[\dfrac{d\gamma}{dx}\right]^2$,

where a term involving $d^2\gamma/dx^2$ has been neglected. Substituting into equation 3.14 we have

$$\left[\frac{d\gamma}{dx}\right]^2 + \frac{2M}{\hbar^2}\left[W - \frac{2Ze^2}{x}\right] = 0.$$

It follows that $\dfrac{d\gamma}{dx} = \sqrt{\left[\dfrac{2M}{\hbar^2}\left(\dfrac{2Ze^2}{x} - W\right)\right]}$

and so $\gamma(x) = \dfrac{1}{\hbar}\sqrt{(4MZe^2)}\displaystyle\int_R^x \sqrt{\left(\dfrac{1}{x} - \dfrac{1}{b}\right)}\,dx,$

where $b = \dfrac{2Ze^2}{T_0}$.

Note that the lower limit of integration has been chosen so that $\gamma(R) = 0$ and the boundary conditions at $x = R$ are as for the rectangular barrier. This integral can be evaluated by a change of variable to θ, where

$$\frac{x}{b} = \cos^2\theta.$$

Thus
$$\gamma(b) = \frac{2b}{\hbar} \sqrt{(2MT_0)} \int_0^{\cos^{-1}\sqrt{(R/b)}} \sin^2\theta \, d\theta$$

$$= \frac{b}{\hbar} \sqrt{(2MT_0)} \left[\cos^{-1}\sqrt{\frac{R}{b}} - \sqrt{\frac{R}{b}}\sqrt{\left(1 - \frac{R}{b}\right)} \right].$$

If now we assume $R \ll b$,

$$\cos^{-1}\sqrt{\frac{R}{b}} = \frac{\pi}{2} - \sin^{-1}\sqrt{\frac{R}{b}} = \frac{\pi}{2} - \sqrt{\frac{R}{b}}.$$

Hence
$$\gamma(b) = \frac{b}{\hbar} \sqrt{(2MT_0)} \left(\frac{\pi}{2} - 2\sqrt{\frac{R}{b}} \right). \qquad\qquad 3.15$$

The discussion can then proceed as in **3.10** with $\gamma(b)$ substituted for $k_2 a$ at the second boundary.

The transmission coefficient (assumed in this case to be equal to the penetration factor, see **3.10**) is therefore given by

$$T = e^{-2\gamma(b)}.$$

3.12 **Nuclear Coulomb barrier**

The nuclear potential barrier must of course be considered in three dimensions. We assume spherical symmetry by taking V, the potential, to be a function only of r, the distance from the nuclear centre. Figure 8 can then be taken to be a section through the three-dimensional barrier.

We can then carry out the same analysis as in the one-dimensional case but Schrödinger's equation takes the three-dimensional form

$$\frac{\partial^2\psi}{\partial x^2} + \frac{\partial^2\psi}{\partial y^2} + \frac{\partial^2\psi}{\partial z^2} + \frac{2M(W-V)}{\hbar^2}\psi = 0.$$

To take advantage of the spherical symmetry of the problem we transform from Cartesian coordinates to r, θ, ϕ, the usual spherical polar coordinates, and Schrödinger's equation then becomes

$$\left[\frac{\partial^2}{\partial r^2} + \frac{2}{r}\frac{\partial}{\partial r} + \frac{1}{r^2}\frac{\partial^2}{\partial\theta^2} + \frac{\cot\theta}{r^2}\frac{\partial}{\partial\theta} + \frac{1}{r^2\sin^2\theta}\frac{\partial^2}{\partial\phi^2} \right]\psi + \frac{2M}{\hbar^2}[W - V(r)]\psi = 0.$$

3.16a

Next, a solution of the form $\psi(r, \theta, \phi) = R(r)Y(\theta, \phi)$ is assumed. When this is substituted into equation **3.16a** we find that all of the terms are either functions of r alone or functions of θ and ϕ. The r-dependent terms are taken to one side of the equation, the terms depending on θ and ϕ to the other side. Each side of the equation must then be a constant independent of r, θ and ϕ. We set each side equal to a separation constant $l(l+1)$. Carrying out this procedure we find that

$$\frac{1}{r^2}\frac{d}{dr}\left[r^2\frac{dR}{dr}\right] + \frac{2M}{\hbar^2}\left[W - V(r) - \frac{l(l+1)\hbar^2}{2Mr^2}\right]R = 0.$$

This constitutes *the radial wave equation.* If a function $G(r) = r\,R(r)$, the so-called *modified radial wave function* be now introduced, the radial wave equation can be written

$$\frac{d^2G}{dr^2} + \frac{2M}{\hbar^2}\left[W - V(r) - \frac{l(l+1)\hbar^2}{2Mr^2}\right]G = 0. \qquad \textbf{3.16b}$$

We note that $|R|^2$ is to be associated with a flux density of particles, that is, the number of particles crossing unit area per unit time. The total number of particles crossing a spherical surface of radius r per unit time will therefore be associated with $4\pi r^2 |R|^2$. This, the particle flux, will be proportional to $|G|^2$. If $l = 0$, we note that equation 3.16b is exactly equivalent to the one-dimensional case analysed above, $G(r)$ playing the role of $\psi(x)$. The particle flux in the spherical case was noted to be proportional to $|G|^2$; in the one-dimensional case it is proportional to $|\psi(x)|^2$. We can therefore, relying on the analogy of G with $\psi(x)$, assume that the penetration factor in the spherical case is given, as in the one-dimensional case, by $e^{-2\gamma}$.

The separation constant was introduced in the somewhat artificial form above in order that l would be equivalent to the angular-momentum quantum number of early quantum theory. $l = 0$ therefore implies that the emitted particle has no angular momentum about the centre of the nucleus, that is, it is emitted radially. We note that if $l \neq 0$, there is an additional term in Schrödinger's equation which effectively adds to the height of the electrostatic potential barrier. This term is said to represent the *centrifugal barrier.* The centrifugal barrier does not depend on electric charge and is effective in all cases, including the case of the emitted particle being a neutron when, because of its electrical neutrality, the Coulomb barrier is not effective.

3.13   **Gamow's theory of alpha decay**

From the above discussion the penetration factor for a nuclear α-particle incident on the Coulomb potential barrier is $e^{-2\gamma}$, where γ is given by equation **3.15.** When the values of the fundamental constants are substituted into this expression, we find that the exponent can be written

$$2\gamma = 3\cdot95\,\frac{Z}{\sqrt{T_0}} - 2\cdot97\sqrt{(ZR)},$$

T_0, the emergent energy of the α-particle, being in millions of electronvolts and R, the nuclear radius, being in fermis.

The penetration factor varies rapidly with α-particle energy. To see how rapid this variation is, we substitute $Z = 90$, with $T_0 = 4\cdot2$ MeV and $R = 10$ fm, the values corresponding to the α-decay of $^{238}_{92}$U. Then $2\gamma = 173\cdot4 - 89\cdot1 = 84\cdot3$ and the penetration factor is therefore $e^{-84\cdot3}$, which is equal to $10^{-36\cdot6}$. Had the α-particle energy been 9 MeV, that is, about double its actual value, then we should have had $2\gamma = 118.5 - 89.1 = 29.4$, in which case the penetration factor would have been $e^{-29.4} = 10^{-12.8}$. Thus for an approximate doubling of α-particle energy there would be a variation of about twenty-four orders of magnitude in the penetration factor.

We now have to consider how many α-particles per second are incident on the potential barrier. This will depend on two factors, namely the number of α-particles existing in the nucleus and the number of collisions each α-particle makes with the potential barrier per unit time.

The number of α-particles existing at any time in a heavy nucleus is probably somewhere between one and ten. These are transient groupings formed by the nucleons in their motion within the nucleus. Fortunately, as we shall see, the decay constant is not critically dependent on this number, which we now denote by P_α. If V_α be the α-particle velocity inside the nuclear potential well, then, the average distance travelled between collisions being of the order of the nuclear radius, the time between collisions will be R/V_α and hence the number of collisions per α-particle per unit time will be V_α/R. If the potential inside the nucleus is close to the value of the potential at infinity, then T_α in the well is equal to T_0, that is, it will be a few million electronvolts in value. V_α is then approximately $c/10 = 3 \times 10^7$ m s^{-1}. We can now write the probability λ that the nucleus will decay per unit time as the product of the number of α-collisions per second with the inside of the barrier and the probability of the α-particle tunnelling through the barrier, as discussed above. Hence

$$\lambda = P_\alpha \frac{V_\alpha}{R} e^{-\{3.95Z/\sqrt{T_0} - 2.97\sqrt{(ZR)}\}}, \qquad\qquad 3.17$$

or, taking logarithms,

$$\log \lambda = \log P_\alpha + \log \frac{V_\alpha}{R} - \left[3.95 \frac{Z}{\sqrt{T_0}} - 2.97\sqrt{(ZR)}\right]\frac{1}{2.3}.$$

The quantities inside the logarithmic terms do not of course affect the result critically. However, T_0 and R, which occur in the remaining terms, clearly have a very large effect on the value of the decay constant.

The theory gives, in a very satisfactory way, the variation of λ with T_0, although this is of different analytical form from the empirical expressions of Geiger and Nuttall.

3.14   **Nuclear unit radius**

As we have discussed in detail above, the energies of scattered α-particles at which departure from Rutherford scattering takes place may be used to deduce nuclear radii. This has now been done for nuclei ranging through the periodic table. It is found that the variation of radius as a function of A is very well described by $R = R_0 A^{\frac{1}{3}}$. We refer to R_0 as the *nuclear unit radius*. Once R_0 is known we can then evaluate the radius of any nucleus. The above theory of α-decay can be used to determine R_0 within close limits.

We can rewrite equation **3.17** in the form

$$\lambda = \lambda_0 \, e^{-(a - b\sqrt{R_0})},$$

regarding a and b as constants to be calculated from the Z, A, T_0 values for any particular α-emitting isotope. The uncertain quantities P_α and V_α are included in λ_0. We assume that these quantities, and hence λ_0, do not change appreciably as we go from one isotope to others with only slightly different Z- and N-values. A determination of λ and T_0 for two different isotopes will then provide two equations from which both λ_0 and R_0 can be found. Because of the occurrence of R_0 in the exponent, this is a very sensitive method for its determination, a variation of 10^{12} in half-life resulting from a change of a factor of two in the value of R_0. The best value of R_0 arrived at by this method is $1 \cdot 48$ fm.

3.15 Fine structure in alpha-particle spectra

It was noted in section 3.3 that α-particles emitted from a single nuclide do not always have the same energy. The spectrum of α-particles from ThC, shown in Figure 10, reveals that in the case of that nuclide there are five possible α-particle energies. If we have regard to equation **3.1** we see that different α-particle energies are only possible (assuming that α-decay is a two-body decay) if different values of Q are possible. These different Q-values in turn demand that the nuclear masses be multivalued. To explain this, the hypothesis is now made that, just as the atomic electron structure can have configurations of different energy content, so may the nucleus have different configurations each with its own associated energy. The lowest energy configuration we call the *ground state*, the others *excited states*. In conformity with mass–energy equivalence, the effective mass of

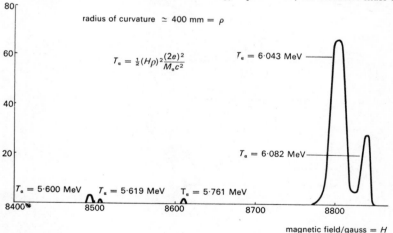

Figure 10 Spectrum of α-particles emitted by ThC (based on measurements by Rutherford and colleagues, 1933)

the nucleus when in an excited state will be the ground-state mass plus E_e/c^2, where E_e is the energy of excitation. Thus the radioactive transformation will have a value of Q which depends on the excited state involved.

There is no evidence for the existence of excited states of the emitted α-particles. However, there are examples of excited states of parent and of daughter nuclei. ThC, quoted above, is a case of the daughter nucleus being created either in the ground state or in an excited state. If the α-particle energies are substituted into equation 3.1, the Q-values may then be calculated. The highest Q-value will correspond to the formation of the ground state. The energies of the excited states, measured from the ground state as zero, will then be given by the amount by which the associated Q-value is less than the ground-state Q-value. The results of this calculation for ThC are shown in Figure 11 in which the states are represented as horizontal lines on a vertical scale of energy, the intervals being proportional to the energies associated with the states. Such a diagram is referred to as an *energy-level diagram*. The excited states may de-excite directly to the ground state, the energies of excitation being carried off by γ-rays (i.e. quanta of electromagnetic energy) or they may 'cascade' through lower states to the ground state with the emission of a series of γ-rays. The time for de-excitation is usually very much less than a nanosecond (10^{-9} s) and thus the

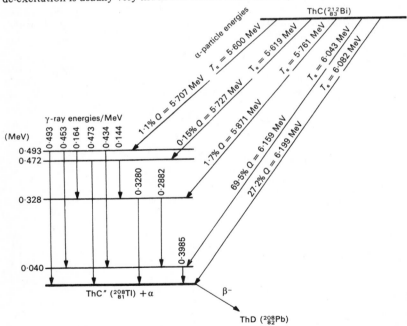

Figure 11 Energy-level diagram of ThC″ showing relationship of -particle and γ-ray spectra.

γ-rays will appear to be in good time coincidence with the α-particles associated with the creation of the excited state. The above interpretation of the decay of ThC can therefore be confirmed by searching for γ-rays, measuring the energy of these γ-rays to compare with the predicted energy from the level diagram and showing that they are in time coincidence with α-particles in the appropriate energy group.

As an example of an excited parent nucleus, we quote ThC′, which has a half life of only 0·3 μs following its formation by β-decay from ThC. It may be formed in its ground state or in one of a set of excited states. The half-life for α-particle emission is so short that α-decay from the excited states is not entirely insignificant compared with photon emission. In a small fraction of cases (approximately 10^{-4} for ThC′) an α-particle of more than normal energy is emitted in this way. Such α-particles are generally referred to as *long-range α-particles*. Compared with the α-particles emitted from the ground state, they carry excess energy of the order of one or two million electronvolts.

3.16 Alpha decay with extremely long half-life

The mass condition of **3.2** has necessarily to be satisfied if α-decay is to be energetically possible. There are instances however where the condition is satisfied but α-decay is not observed. This is always where the mass difference, and hence the energy released in the α-decay transformation, is very small. The α-particle, having low energy, has to penetrate a very wide barrier. As a consequence λ is small and $T_{\frac{1}{2}}$ is very long even if the α-particle has no angular momentum with respect to the centre of the daughter nucleus. If, because of angular-momentum conservation requirements, the α-particle has angular momentum relative to the daughter nucleus, then the centrifugal barrier will further decrease the decay probability. As a result the activity per unit mass of sample of material will be very small.

There is a limit to the mass of sample which can be under examination by a given detection system at any one time. Further, any detection system has a background counting rate due to cosmic rays and radioactive contamination in its structural materials and its surroundings. This background counting rate arising from random effects has statistical fluctuations which set a limit to the accuracy with which an activity can be measured. In the extreme case, it may be impossible to distinguish between a very weak activity and no activity.

For this reason, nuclei once believed to be stable may subsequently be found to be unstable. For example ^{142}Ce appeared, until recently, in nuclear data tables as a stable isotope. There is now evidence that it is α-unstable, emitting α-particles of about 1·5 MeV and having a half-life of 5×10^{15} years.

3.17 Summary

The use of the α-particle as a charge probe for the measurement of the electrostatic field within the atom led to the concept of the potential barrier and to an estimate of nuclear size. It also revealed a paradox with respect to the emission of α-particles with insufficient energy to have surmounted the potential barrier. The resolution of this paradox by the abandonment of classical dynamics in favour of wave mechanics led to an explanation of the observed relationship between half-life and α-particle energy. The measurements of these two quantities for two α-emitting nuclides were used to determine R_0, the nuclear unit radius which enters the formula for the nuclear radius, namely $R = R_0 A^{\frac{1}{3}}$. The interpretation of the fine structure observed in the energy spectra of α-particles established the existence of excited states of the nucleus and led to the introduction of energy-level diagrams.

Chapter 4
Radioactivity: Beta Decay

4.1   **Introduction**

Beta decay, the most generally occurring mode of radioactive transformation, takes place between neighbouring isobars (i.e. without change in A and with a change of one in Z). In contrast to α-decay, which is a phenomenon limited to nuclei with medium and high A-values, β-decay has been observed for nuclei with all A-values from one upwards. Essentially in β-decay a neutron switches into a proton or vice versa. When the switch occurs a β-particle, of negative sign of electric charge if a neutron switch, of positive sign if a proton has switched, is observed to be emitted. Careful experimentation has failed to reveal any difference between the physical properties of the negative β-particle and those of the electron of atomic structure, and we assume that these particles are identical. The positive β-particle, apart from the sign of its electric charge, has the same properties as the negative β-particle. The β-particles are sometimes named *negatron* (perhaps more properly, but less usually, *negaton*) and *positron* (or *positon*), *electron* then being available to apply generically to either.

4.2   **Beta decay and the conservation laws**

The measurement, by Chadwick in 1914, of the energy of β-particles emitted from a source containing a single isotopic species revealed a continuous spectrum of energy ranging from zero to a finite maximum value. If it is assumed that, as in α-decay, the parent and daughter nuclei have well-defined mass values, then the conservation of mass–energy and linear momentum requires that there be at least three 'products' of the decay, that is, one product in addition to the β-particle and the recoiling daughter nucleus. Careful measurement of the energy absorbed in massive calorimeters containing strong β-sources indicated an energy per decay corresponding to the mean β-energy, not to the maximum β-energy. Thus the third 'product', if such existed, did not deposit any energy in the material of the calorimeter (Ellis and Wooster, 1927).

In addition to the difficulty thus presented in respect of energy conservation, β-decay set a problem with respect to conservation of angular momentum. The simplest β-emitter is the free neutron, which, with a half-life of about thirteen minutes, decays to a proton. We start with a neutron which has intrinsic angular momentum of $\frac{1}{2}\hbar$. If we end with only a proton and electron, each having intrinsic angular momentum of $\frac{1}{2}\hbar$ and only permitted by the rules of quantum mechanics to have an angular momentum of relative motion in multiples of $\hbar$, then clearly angular momentum cannot be conserved. The question of conservation of linear momentum was also raised by the existence of cloud-chamber photographs purporting to show the decay of ^{6}He in which the direction of the recoiling daughter nucleus was not accurately collinear with the direction of the emitted β-particle.

4.3 **The neutrino hypothesis**

The three conservation laws, those of energy, linear mome tum and angular momentum, can be satisfied in β-decay by the simple expedient of postulating (as was done by Pauli in 1933) the emission of a neutral particle, the *neutrino*, in addition to the β-particle. We must assign to this new particle the property of having practically no interaction with matter, in order to explain the failure to detect such a particle in the early calorimeter experiments. More recently, the interaction of the neutrino with matter has been observed in elaborate experiments with large baths of scintillating liquid in which neutrinos from the β-emitting products in a reactor undergo the reaction

$$p + \bar{\nu} \rightarrow n + e^+,$$

where $\bar{\nu}$ represents the neutrino (Reines and Cowan, in 1959). Thus the early philosophical objections to the acceptance of the existence of a particle whose detection was virtually impossible by definition, have now been removed.

The neutron-to-proton switch, which, as mentioned above, occurs as the β-decay of free neutrons and has been observed to take place in neutron beams emerging from reactors, can be described by

$$n \rightarrow p + e^- + \bar{\nu}. \tag{4.1}$$

Free protons on the other hand are stable against β-decay. However, in the conditions existing inside the nucleus, it is assumed that the reaction analogous to **4.1** takes place, namely

$$p \rightarrow n + e^+ + \nu, \tag{4.2}$$

giving rise to positron emission.

It is believed that the neutrinos involved in **4.1** and **4.2** are not identical. ν, associated with positron decay, is called the neutrino; $\bar{\nu}$, associated with negatron decay, is called the antineutrino. Neutrinos, electrons and μ-mesons constitute a family of particles known as *leptons*. If we regard the electron and positron as a particle and antiparticle pair, we note that in β-decay the two leptons appearing are always in a particle and antiparticle combination.

The neutrino, for reasons which will be discussed later, is believed to have zero rest mass and always to have a velocity equal to the velocity of light. The conservation of angular momentum in β-decay requires that the neutrino should have the same spin angular momentum as the nucleons and the electrons, namely $\frac{1}{2}\hbar$.

4.4 Mass-difference conditions necessary for beta decay

In order that β-decay be energetically possible, a certain inequality must be satisfied by the masses of the parent and daughter isotopes.

4.4.1 β^- decay

Let $_Z^A X$ represent a nuclide which is unstable against β^- decay and let $_{Z+1}^A Y$ represent the isobar into which it decays. Then the process starts from X with its complete complement of orbital electrons and, it is assumed, with the electrons in their ground-state configuration. Following the emission of the β^- particle, which it is assumed does not interact with the electron system in its passage out through the atom, the daughter Y will have a nuclear charge $Z + 1$ but only the electron complement of X, namely that corresponding to a nuclear charge Z. There will be slight adjustments in orbitals necessary by virtue of the change in Z; Y will also be singly ionized. If now M_X is the mass of the X-atom and M_Y^+ the mass of the singly ionized Y-atom, the mass–energy equation of the reaction is

$$M_X = M_Y^+ + m_e + Q, \qquad\qquad 4.3$$

where Q is the total kinetic energy available for the electron, neutrino and daughter atom, m_e is the mass of an electron, and the mass of the neutrino is assumed to be zero.

Now the (first) ionization energy of the Y-atom will be given by

$$I = M_Y^+ + m_e - M_Y,$$

where M_Y is the mass of the Y-atom in its ground state. Equation 4.3 can then be written

$$M_X = M_Y + I + Q.$$

I will be of the order of electronvolts and will usually be negligible compared to the reaction energy. On this assumption the condition that β^- decay be energetically possible, namely that $Q > 0$, leads to the requirement

$$M_X > M_Y. \qquad\qquad 4.4$$

4.4.2 β^+ decay

In the case of β^+ decay the roles of X and Y are reversed. Y is now the parent nucleus. The daughter will have one electron over its full complement when it is formed. Depending on the atom concerned, this electron may remain attached, in which case a negative ion results, or it may dissociate from the atom leaving a neutral atom and a free electron. The energy difference between these two possibilities (i.e. the electron-attachment energy) is however only a few electronvolts and will usually be negligible compared to the reaction energy. In the final state, in addition to the daughter atom, we thus have one free electron and one positron. The mass–energy equation is therefore

$$M_Y = M_X + 2m_e + Q.$$

For $Q > 0$, therefore,

$$M_Y > M_X + 2m_e.\qquad\qquad 4.5$$

4.4.3 *Electron capture*

Having regard to the inequalities **4.4** and **4.5**, the question arises of a possible stability of neighbouring isobars in the event of

$$M_X + 2m_e > M_Y > M_X.$$

Although β-decay, as it has been discussed above, is not energetically possible, a process equivalent to positron emission is possible and is observed to take place.

The reaction

$$p + e^- \rightarrow n + \nu\qquad\qquad 4.6$$

is equivalent to reaction **4.2** in respect of the daughter nucleus formed. However, the interaction between a nucleon (i.e. proton or neutron) and a lepton is the so-called *weak interaction*, which we discuss later, and is of a magnitude which makes the reaction **4.6** experimentally unobservable in terms of a free-proton target bombarded by a beam of electrons. In the atom, the nucleus in a sense is being constantly bombarded by those orbital electrons whose wave functions have finite amplitudes within the nuclear volume. Thus, despite the weakness of the interaction process, **4.6** can be expected to occur in the course of time. This process in general terms is known as *electron capture*. The two electrons in the K-shell have, compared with electrons in other orbitals, a relatively large probability of being found within the nuclear volume and the process therefore usually proceeds by the capture of one of these electrons. In that event, it is referred to as 'K-*capture*' to distinguish it from an event involving the capture of an L- or even an M-shell electron. The process of capture from shells other than the K-shell has a smaller but still finite probability and is observed.

The end product of electron capture in terms of emitted particles is solely a neutrino. It should be noted that, since the process in this case is a two-body process, as distinct from the three-body process involving β-particle emission, the neutrinos from a given nuclide are monoenergetic. It is not usually possible to detect these neutrinos. However, the disappearance of the electron leaves a vacancy in an atomic shell. The filling of this vacancy by an electron jumping in from another shell will result either in the emission of an X-ray or an Auger electron, both of which can be detected with high efficiency. For example if the electron vacancy is in the K-shell, then it can be filled by an electron from the L-shell jumping in with the emission of a K X-ray and the creation of a vacancy in the L-shell. (In this summary account we are neglecting the multiplicity of the L-shell, which has in fact three components L_I, L_{II}, L_{III}.) alternatively, the electron from the L-shell may, as before, fill the K-shell vacancy but instead of the emission of the energy difference in the K- and L-shell binding energies, namely $B_K - B_L$, in the form of an X-ray, this energy may concentrate on a second L-shell electron, ejecting it from the atom as an Auger electron having an energy of $B_K - 2B_L$. In this event two vacancies are left in the L-shell, which are then filled by similar processes by electrons jumping from the outer shells. Thus the vacancies move out through the shells till finally an ion is left which will then neutralize itself by finding free electrons, if such are available in its environment.

As the final state immediately following electron capture consists of the daughter nucleus with one vacancy in a certain shell, but with its full electron complement, the mass–energy relation is

$$M_Y = M_X + E' + Q,$$

where E' is the energy necessary to produce the vacancy by promoting an electron to an outer orbit. In principle E' need only be a few electronvolts in magnitude if an electron in an outer shell is captured. Even when the capture is from a deeper shell, E' will in most cases be negligible in comparison with the reaction energy. Thus the condition that electron capture be energetically possible is

$$M_Y > M_X. \tag{4.7}$$

It should be noted that if the inequality **4.5** is satisfied then **4.7** is also satisfied. In that event the processes of positron emission and electron capture occur in competition.

Inequality **4.7** taken with **4.3** means that one of two neighbouring isobars must in all circumstances be unstable with respect to decay into the other. In Figure 12 the three modes of β-decay are related to the mass differences.

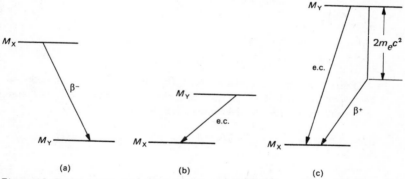

Figure 12 Relationships between mass values of neighbouring isobars giving rise to (a) β^- decay, (b) electron capture only, (c) competition between electron capture and β^+ decay

4.5 **Decay energy and beta-particle energy**

We define the decay energy to be the difference between the ground-state masses of the parent and daughter nuclei. If the transition is from ground state to ground state, then the whole of the decay energy is available as kinetic energy to the products of the decay in the case of β^- decay and electron capture. In the case of

β^+ decay, the decay energy less $1 \cdot 022$ MeV (i.e. twice the electron rest mass) is available. Very frequently, however, the transition is not to the ground state of the daughter nucleus but to an excited state which quickly decays by γ-ray emission to the ground state. In that event the energy of excitation of the excited state has to be subtracted from the decay energy to arrive at the energy available to the particles, as discussed above.

4.6   The energy distribution among products in beta decay

The energy available in a β-decay transition is shared by the daughter nucleus, the β-particle and the neutrino, the division being governed among other factors by the requirement that linear momentum be conserved. As there are three bodies involved, the division can take place in an infinite number of ways, and therefore each product has an energy lying in a continuous range from zero to a fixed maximum value for a particular transition.

The daughter nucleus will have its maximum possible recoil energy when the neutrino energy is zero, the β-particle energy has its maximum value for the transition in question and the linear momentum of the daughter nucleus balances that of the β-particle. The total energy E_β of the β-particle is given by

$$E_\beta = T_\beta + m_e c^2.$$

where T_β is its kinetic energy; also from the special theory of relativity, the β-particle momentum p_β is given by

$$p_\beta^2 c^2 = E_\beta^2 - m_e^2 c^4.$$

It therefore follows that

$$p_\beta^2 c^2 = T_\beta^2 + 2T_\beta m_e c^2,$$ **4.9**

If now p_D is the linear momentum of the daughter nucleus, whose kinetic energy will be small compared to its rest mass and whose motion can therefore be adequately described by Newtonian dynamics, then we have

$$2M_D T_D = p_D^2 = p_\beta^2 = \frac{T_\beta^2}{c^2} + 2T_\beta m_e.$$

Therefore $$T_D = \frac{T_\beta^2}{2M_D c^2} + T_\beta \frac{m_e}{M_D}.$$ **4.10**

If this result be now applied to the particular case of ^{6}He, which has a comparatively low mass value for a β-emitter and a comparatively large maximum β-particle energy of $3 \cdot 5$ MeV and which should therefore have a relatively large recoil energy, we find $T_D = 1 \cdot 4$ keV. Thus less than one part in two thousand of the available energy is taken away by the recoil. In the case of other β-emitters where a heavier recoil mass and a lower β-particle maximum energy are involved the fraction will be even smaller. Consequently one can, in most circumstances, equate the decay energy to the maximum β-particle energy, allowance having been made for any subsequent γ-ray emission as discussed above.

4.7 Range of values of $T_{\beta,\,max}$

There are several hundred nuclides classified as β-emitters. They range in A-value from $A = 1$ for the neutron to the isobars with $A = 256$ (e.g. $^{256}_{99}Es$).

The largest value of decay energy in a β-transition is 18 MeV, which occurs in the case of the nuclide $^{8}_{5}B$. However, this positron emitter decays to an excited state of $^{8}_{4}Be$ and the maximum particle energy is about 14 MeV. $^{12}_{7}N$, which has a slightly smaller decay energy of 17·6 MeV, decays to the ground state of $^{12}_{6}C$ and consequently the maximum positron energy is 16·6 MeV. At the other end of the scale cases are known of decay energies considerably less than 10 keV. The value of $T_{\beta,\,max}$ measured in the case of ^{187}Re, in what is believed to be a ground-state–ground-state transition, is about 2 keV. When the decay energy is as small as this then the assumptions made in section 4.4, concerning the effect of the orbital electrons in the atoms, have to be considered carefully. A reorganization of the extranuclear electrons may involve in the case of heavy atoms some 12 keV. In β⁻ decay, energy is provided by this reorganization; in β⁺ (or electron capture) it has to be provided out of the nuclear energy store. A β⁻-active heavy atom for which, in the neutral state, the decay energy is, say, 10 keV would be stable if completely ionized. The range of particle energy in β-decay from a few thousand electronvolts to about 17 MeV is very much wider than the range of particle energies involved in α-decay, which is approximately 2 MeV to 9·2 MeV.

4.8 Range of half-lives in beta decay

In the case of β-decay, as for α-decay, the more energetic the emitted particle the shorter, as a general rule, is the half-life of the nuclide. For example, $^{12}_{7}N$, which emits a very-high-energy positron, has a half-life of 11 ms. On the other hand, ^{187}R which emits a 2 keV β⁻ particle, has a half-life of 4×10^{10} years. An even longer half-life, namely $1\cdot1 \times 10^{11}$ years, has been measured for the β⁻ emitter ^{138}La. This range of half-lives is not so great as the range for α-decay, where the lower limit is less than a microsecond and the upper limit, probably set by experimental technique, is at least 2×10^{17} years.

An early empirical attempt to relate the half-life to the maximum energy of β-particle of naturally occurring β-emitters was made by plotting $\log \lambda$ against $\log E_{max}$. The resulting plot is called a *Sargent diagram*. On this diagram the β-emitters were found to fall into two distinct groups. Those with the shorter half-life for a given E_{max} were said to involve 'allowed' transitions, those with the longer half-life to involve 'forbidden' transitions.

4.9 Fermi theory of beta decay

Two experimental features of β-decay present obvious challenges for theoretical interpretation, namely the shape of the measured energy spectrum of emitted β-particles and the relation between half-life and maximum β-particle energy. A theory was devised by Fermi (1934) to provide an interpretation of these

57 Fermi theory of beta decay

features. This theory has largely stood the test of time and is still a convenient basis for the theoretical discussion of β-decay.

An analogy is made between β-decay and the transition between two states of an atom with the emission of electromagnetic radiation. The β-particle and neutrino, like the electromagnetic quantum in the atomic case, do not exist prior to the transition taking place, their creation and emission forming part of the transition process.

In the atomic case, the transition rate between initial and final states is given by perturbation theory (see, for example, R. M. Eisberg, *Fundamentals of Modern Physics*, Wiley, 1961, p. 268) and can be written as

$$\frac{2\pi}{\hbar} \, v_{fi}^* \, v_{fi} \, \rho_f,$$

where ρ_f is the number of final energy states per unit energy interval and v_{fi} is the *matrix element* of the interaction potential V, the perturbation causing the transition. In this theory, v_{fi} is defined by the equation

$$v_{fi} = \int \psi_f^* \, V \psi_i \, d\tau,$$

where ψ_i and ψ_f are the eigenfunctions (see section 6.4) describing the initial and final states of the system. In the atomic case, V represents the potential energy of the electric charges and magnetic moments of the atomic system in the perturbing electromagnetic field.

In the case of β-decay, a new interaction has to be assumed to provide the perturbation and initially the simple assumption is made that V can be replaced by a constant g which is called the *β-decay coupling constant*. With this assumption we now write

$$v_{fi} = g \int \psi_f^* \, \psi_i \, d\tau.$$

Nov the initial system is simply, let us say, the β-unstable nucleus (Z, A) and therefore we will write $\psi_i = \psi_{Z,A}$.

The final system consists of the daughter nucleus $(Z \pm 1, A)$ (the sign depending on whether we have β^- or β^+ decay) together with the outgoing β-particle and neutrino. The eigenfunction ψ_f is therefore to be obtained by multiplying the eigenfunctions of the three decay products; thus

$$\psi_f = \psi_{Z \pm 1, A} \, \psi_\beta \, \psi_\nu.$$

The electron and neutrino we shall at this stage in the discussion assume to be adequately described by free-particle plane waves, that is, we assume them to emerge through a field-free region. Both of these particles have momenta in a range in which the associated de Broglie wavelength is very much larger than nuclear dimensions. Consequently as a simplification we take ψ_β and ψ_ν to be constants independent of the space coordinates throughout the nuclear volume. For the purpose of specifying boundary conditions we assume that the whole system is enclosed in a box of dimension L. Taking ψ_β, ψ_ν both equal to $L^{-\frac{3}{2}}$

normalizes the probability of finding β-particle and neutrino in the box to unity. We can therefore write

$$\psi_f = L^{-3} \, \psi_{Z \pm 1, A},$$

and it follows that

$$v_{fi} = gL^{-3} \int \psi^*_{Z \pm 1, A} \, \psi_{Z, A} \, d\tau = gL^{-3}M. \qquad \textbf{4.11}$$

M is termed the *nuclear matrix element* and it is to be noted that under the assumptions made it is independent of the β-particle and neutrino momenta.

It remains to discuss ρ_f, the density of final states. We start from a result of non-relativistic quantum mechanics, that if an electron of kinetic energy T_β is confined in a box of dimension L, then $N(T_\beta)$, the density of final states, is given by

$$N(T_\beta) \, dt_\beta = \frac{m^{\frac{3}{2}} L^3 T_\beta^{\frac{1}{2}} \, dT_\beta}{2^{\frac{1}{2}} \pi^2 \hbar^3}.$$

We now proceed to express this in terms of momentum. Using the non-relativistic relation

$$\frac{p_\beta^2}{2m} = T_\beta,$$

we have $\quad \dfrac{p_\beta \, dp_\beta}{m} = dT_\beta.$

Thus $\quad N(T_\beta) \, dT_\beta = \dfrac{L^3 p_\beta^2 \, dp_\beta}{2\pi^2 \hbar^3}. \qquad \textbf{4.12}$

It can be shown that this relation is also correct in the relativistic case and, as it now stands, it may be applied to β-particles and neutrinos. We now proceed to apply the equation 4.12 to the β-decay discussion. We consider the energy of the final state to lie between E_f and $E_f + dE_f$. If ρ_f is the density of final states, then the number of states in the energy range dE_f is $\rho_f \, dE_f$. Corresponding to dE_f there will be a range dE_β of β-particle energies and a range dE_ν of neutrino energies.

It follows from equation 4.12 that there are

$$N(E_\beta) \, dE_\beta = \frac{L^3 p_\beta^2 \, dp_\beta}{2\pi^2 \hbar^3}$$

β-particle states and

$$N(E_\nu) \, dE_\nu = \frac{L^3 p_\nu^2 \, dp_\nu}{2\pi^2 \hbar^3}$$

neutrino states. The number of states of the system is then obtained by forming the product of these two expressions relating to the individual components. Thus

$$\rho_f \, dE_f = \frac{L^6 p_\beta^2 p_\nu^2 \, dp_\beta \, dp_\nu}{4\pi^4 \hbar^6}.$$

We now consider the β-particle energy to be specified and associate dE_f solely with dE_v. We therefore write

$$\rho_f = \frac{L^6 p_\beta^2 p_v^2 \, dp_\beta}{4\pi^4 \hbar^6} \left[\frac{dE_f}{dp_v}\right]^{-1}.$$

Assuming the neutrino rest mass to be zero we have $E_v = p_v c$. Therefore

$$\frac{dE_v}{dp_v} = c.$$

Also $E_\beta + E_v = E_{\beta,max}$,

where $E_{\beta,max} = m_0 c^2 + T_{\beta,max}$.

In these expressions $E_{\beta,max}$ and $T_{\beta,max}$ are the total and kinetic energies respectively of the β-particle at the upper limit of the energy spectrum. Therefore

$$p_v = \frac{E_{\beta,max} - E_\beta}{c} = \frac{T_{\beta,max} - T_\beta}{c}.$$

We can then write

$$\rho_f = \frac{L^6 p_\beta^2 (T_{\beta,max} - T_\beta)^2 \, dp_\beta}{c^3 4\pi^4 \hbar^6}. \qquad \textbf{4.13}$$

Substitution from equations **4.11** and **4.13** into the expression for the transition rate then yields

$$P(p_\beta) \, dp_\beta = \frac{g^2 M^* M}{2\pi^3 \hbar^7 c^3} (T_{\beta,max} - T_\beta)^2 p_\beta^2 \, dp_\beta, \qquad \textbf{4.14}$$

where $P(p_\beta) \, dp_\beta$ is the probability per unit time that a β-particle of momentum in the range p_β to $p_\beta + dp_\beta$ will be emitted, i.e. $P(p_\beta)$ is the spectral function of the β-particle momentum spectrum. As it is particle momentum which is directly measured in a magnet spectrometer, and as the most accurate spectra have been measured with such an instrument, it is convenient to leave the spectral function in this form rather than to convert entirely to energy. Note that the spectral function separates into three factors, the first involving the universal constants $\hbar, c, g$, the second depending on the nuclear matrix element M, assumed independent of the lepton momenta, and the third, the *statistical factor*, giving the spectral shape.

4.10 Beta-particle momentum spectrum

The theoretical spectrum given by equation **4.14** is a bell-shaped curve of the same general shape as a typical measured spectrum. At the low-momentum end of the distribution we may obtain an approximation to the predicted shape by neglecting all terms of higher order than p_β^2 and we see that the distribution should be proportional to p_β^2, i.e. should be parabolic. Since, up to this point, electric charge has been assigned no role in the process, the theoretical spectral shape is the same for positrons and electrons. When, however, measured spectra are compared with the theoretical spectra adjusted to fit at the maximum values, it is found that for low momentum values the measured spectra for β^- particles lie consistently above the theoretical spectra and are almost linear in shape in the neighbourhood of the origin. On the other hand, for β^+ particles the measured spectra lie consistently below the theoretical curves.

An explanation of the departure from the theoretical distributions derived above is to be looked for in the assumption in the theory that the outgoing electron can be treated as a free-particle plane wave. This assumption neglects the fact that there is an interaction between the electron's charge and that of the daughter nucleus. The Coulomb force between these particles will decelerate the outgoing particle, if it is a negatron, and thus increase the proportion of low-momentum particles in the spectrum; in the case of positrons the Coulomb force will accelerate the positrons and reduce the relative number of low-momentum particles.

A more exact treatment requires the substitution for ψ_β in section 4.9 of an eigenfunction which will take the Coulomb interaction into account, and which will have a greater amplitude in the region near the nucleus, in the case of negatrons, and a smaller amplitude in that region, in the case of positrons, than the plane-wave amplitudes. This substitution leads to the introduction into the right-hand side of equation **4.14** of a factor $F(Z, p_\beta)$, the *nuclear Coulomb factor*, which in the completely relativistic treatment is of a complicated form. It has been tabulated as a function of Z and p_β (I. Feister, *Physical Review*, 1950). A simplified non-relativistic treatment leads to the following expression for $F(Z, p_\beta)$ which for many purposes is a close enough approximation to the tabulated value.

$$F(Z, p_\beta) = \frac{2\pi\delta}{1 - e^{-2\pi\delta}},$$ **4.15**

where $\delta = \pm Z\alpha \dfrac{E_{\beta,\max}}{cp_\beta}.$

α is the fine structure constant, $1/137$, and the positive sign is to be taken for β^- decay, the negative sign for β^+ decay.

For β^- decay and low values of momentum p_β, δ is large and positive and therefore

$$F(Z, p_\beta) \simeq 2\pi Z\alpha \frac{E_{\beta,\max}}{cp_\beta} \propto (p_\beta)^{-1}.$$

Taken with the p_β^2 dependence of the remaining factors in equation **4.14**, this leads to an overall dependence on p_β in agreement with the linear rise of the experimentally measured spectra.

For β^+ decay and low values of momentum, δ is large and negative. Then the denominator of equation **4.15** is dominated by the exponential term. Hence

$$F(Z, p_\beta) \simeq 2\pi |\delta| e^{-2\pi|\delta|}.$$

The presence of the exponential has the effect of severely reducing the relative number of low-momentum positrons, in accord with the experimental observations.

The nuclear Coulomb factor must of course approach unity as $Z \to 0$, its effect becoming increasingly more marked as Z increases. The full quantum-mechanical treatment indicates that it not only redistributes the particles in the spectrum, as the naïve argument based on the accelerating effect on the outgoing particle might suggest, but that it increases the probability of β-emission throughout the spectrum for negatrons, thus increasing the overall probability of β^- decay. It has the opposite effect in the case of positron emission.

4.11 The Kurie plot

$P(p_\beta)$ in equation **4.14** corresponds to the number of particles in the momentum range p_β to $p_\beta + dp_\beta$ to be found in an experimental measurement of the spectrum. A direct plot of $P(p_\beta)$ against p_β yields, as we saw above, a bell-shaped curve. When experimental errors are involved, it is not a simple matter to make a detailed comparison of observations with the theoretical predictions. Further the determination of the end point of the spectrum, from which the important quantity $T_{\beta,\max}$ is to be evaluated, is a difficult exercise because the approach to the momentum axis is predicted to be parabolic and of course the particle numbers in each channel of increasing momentum are approaching zero. In the experimental situation therefore these points have a low statistical accuracy and are merging into a statistically fluctuating background.

If, however, use is made of the theoretically predicted shape of the spectrum and the quantity

$$\left[\frac{P(p_\beta)}{p_\beta^2 \, F(Z, p_\beta)} \right]^{\frac{1}{2}}$$

is computed for each momentum range, then this quantity is predicted by equation **4.14** to be equal to a constant multiplied by $T_{\beta,\max} - T_\beta$. Therefore if the expression is plotted against T_β rather than against p_β, two important consequences follow. Firstly the degree of linearity of the plot establishes how well equation **4.14** predicts the spectral shape. Secondly if the plot is linear then a simple extrapolation of the plot to cut the T_β axis gives a value of $T_{\beta,\max}$ with a statistical accuracy determined largely by the good statistical accuracy of the points near the maximum in the bell-shaped spectrum.

This useful form of presentation, exemplified in Figure 13, is termed a *Kurie plot*. Many examples are known of spectra leading to very linear Kurie plots. Where there is a departure from linearity (after all allowance has been made for finite source thickness and scattering from materials behind the source, both effects which can cause distortion of the low-energy part of the spectrum), then it is taken as an indication that v_{fi} for the nuclide in question is not independent of p_β.

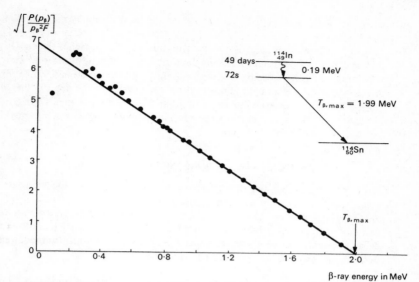

Figure 13 Kurie plot of the β^- spectrum of ^{114}In. $P(p_\beta)$ denotes the number of particles in a constant interval of momentum, while the kinetic energy plotted along the horizontal axis corresponds to the midpoint of momentum interval

4.12 Separation of complex beta spectra

There are many examples of β-transitions in which the residual nucleus may be left in its ground state or may be left in one of a set of excited states. This is analogous to fine structure in the case of α-decay. When there is, let us say, one excited state involved, there will be two β end-point energies and two superimposed continuous β-spectra. These spectra may be separated by making a Kurie plot which in the region beyond the end point of the lower-energy spectrum will be linear. By extrapolating this linear plot back to low momenta, the β-particle spectrum associated with the excited state can then be constructed by subtraction. Where there are two or more excited states involved, then the same process can be carried through in successive steps.

4.13 The mass of the neutrino

It was assumed in the derivation of equation **4.14** that the neutrino had zero rest mass. If we do not make this assumption then the full relationship

$$E_\nu^2 = p_\nu^2 c^2 + m_\nu^2 c^4$$

must be used. When we follow through the consequences of this, it is found that the effect is to modify the spectral distribution at the high energy tip, leading to a sharper cut-off to a value of $T_{\beta,\max}$ less than the value found by extrapolating the Kurie plot.

Thus the careful investigation of the Kurie plot as it approaches the energy axis is an accepted way of establishing the neutrino mass. No departure from linearity in this region has been established with certainty and the experiments therefore lead to the setting of an upper limit to the possible value of the neutrino mass. On this basis it can now be said to be less than 250 eV, i.e. less than 1/2000 of the electron mass and the experimental results are compatible with it being equal to zero.

4.14   The theoretical half-life and comparative half-life of beta emitters

$P(p_\beta)\, dp_\beta$ is the probability that β-decay will take place and that the β-particle will have a value of momentum in a particular momentum range. Thus $d\lambda = P(p_\beta)\, dp_\beta$ may be defined as the partial decay constant associated with a particular momentum requirement placed on the outgoing β-particle. To find the decay constant in the usual sense, we have to remove this momentum requirement by integrating over all values of β-particle momenta. Hence

$$\lambda = \int_0^{p_{\max}} \frac{g^2 M^* M}{2\pi^3 c^3 \hbar^7}\, p_\beta^2 (T_{\beta,\max} - T_\beta)^2\, F(Z, p_\beta)\, dp_\beta$$

$$= \frac{g^2 M^* M}{2\pi^3 \hbar^7}\, m_0^5\, c^4 f(Z, E_{\beta,\max}), \qquad\qquad 4.16$$

where

$$f(Z, E_{\beta,\max}) = \frac{1}{m_0^5 c^7} \int_0^{p_{\max}} F(Z, p_\beta) p_\beta^2 (T_{\beta,\max} - T_\beta)^2\, dp_\beta. \qquad 4.17$$

Since $E_\beta = m_0 c^2 + T_\beta$, with a similar expression for $E_{\beta,\max}$, we may write equation 4.17 in the dimensionless form

$$f(Z, E_{\beta,\max}) = \int_0^{p_{\max}} F(Z, p_\beta) \left[\frac{E_{\beta,\max} - E_\beta}{m_0 c^2} \right]^2 \left[\frac{p_\beta}{m_0 c} \right]^2 \frac{dp_\beta}{m_0 c}.$$

An analytic expression for this integral cannot be given. It has been numerically computed and its value, for given values of Z and $E_{\beta,\max}$, tabulated and graphed (Feenberg, E. and Trigg, G., *Reviews of Modern Physics*, vol. 22, 1950, p. 399). However, the analysis can be taken further in the particular case where $F(Z, p_\beta)$ can be taken to be unity. This will correspond to β-transitions where Z has small values, strictly speaking where $Z = 0$. Then

$$f(0, E_{\beta,\max}) = \int_0^{p_{\max}} \left[\frac{E_{\beta,\max} - E_\beta}{m_0 c^2} \right]^2 \left[\frac{p_\beta}{m_0 c} \right]^2 \frac{dp_\beta}{m_0 c}$$

$$= \left[\frac{E_{\beta,\max}}{m_0 c^2} \right]^5$$

$$\int_{m_0 c^2}^{E_{\beta,\max}} \left[1 - \frac{E_\beta}{E_{\beta,\max}} \right]^2 \left\{ \left[\frac{E_\beta}{E_{\beta,\max}} \right]^2 - \left[\frac{m_0^2 c^4}{E_{\beta,\max}^2} \right] \right\}^{\frac{1}{2}} \frac{E_\beta}{E_{\beta,\max}} \frac{dE_\beta}{E_{\beta,\max}},$$

using $p_\beta^2 c^2 = E_\beta^2 - m_0^2 c^4$.

Let $\dfrac{E_\beta}{E_{\beta,\max}} = x$ and $\dfrac{m_0 c^2}{E_{\beta,\max}} = a$.

Then $f(0, E_{\beta,\max}) = \left[\dfrac{E_{\beta,\max}}{m_0 c^2} \right]^5 \displaystyle\int_a^1 (1-x)^2 (x^2 - a^2)^{\frac{1}{2}} x \, dx$.

If now a can be assumed small (i.e. $E_{\beta,\max} \gg m_0 c^2$) then

$$f(0, E_{\beta,\max}) = \left[\frac{E_{\beta,\max}}{m_0 c^2} \right]^5 \int_0^1 (1-x)^2 x^2 \, dx = \frac{1}{30} \left[\frac{E_{\beta,\max}}{m_0 c^2} \right]^5$$

Thus $\lambda = \text{constant} \times (E_{\beta,\max})^5$.

This result has been derived for low-Z nuclei emitting high-energy β-particles but its prediction that λ should be strongly dependent on the maximum energy of the emitted particle is true in the general case. In a comparison of decay constants (or equivalently in a comparison of the half-lives) of two isotopes it is very frequently convenient to remove the very strong effect of β-particle kinematics by forming the product

$$f(Z, E_{\beta,\max}) T_{\frac{1}{2}},$$

which is called the *comparative half-life* and often referred to as the '*ft*' value. From equation 4.16 we see that

$$f(Z, E_{\beta,\max}) T_{\frac{1}{2}} = \frac{f(Z, E_{\beta,\max})}{\lambda} \ln 2 = \frac{\text{constant}}{M^*M}. \qquad \textbf{4.18}$$

A measurement of the '*ft*' value thus gives a measure of the nuclear matrix element for the transition.

A survey of the measured '*ft*' values for a range of β-emitters reveals that, far from being constant, they vary over many orders of magnitude. As a consequence it is frequently convenient to work with log *ft*. Measuring *ft* in seconds, then log *ft* values range from about 3·5 to 23. Since the greatest possible value of M^*M will be unity and will arise when the normalized eigenfunctions $\psi_{Z \pm 1, A}$

and $\psi_{Z,A}$ are identical, we assume that the eigenfunctions are closest to being identical in the cases that give rise to the lowest observed ft values. On this assumption substituting $ft = 10^{3.5}$ s and $M^*M = 1$ in equation 4.18 and using the known values of the other constants involved, we arrive at $g = 1.4 \times 10^{-62}$ J m^3 for the value of the β-decay coupling constant as introduced above.

The cases involving ft values of the order of $10^{3.5}$ s arising from the eigenfunctions of the initial and final states being very similar, are referred to as *allowed transitions*. The so called *mirror nuclei* of which $^{17}_{8}O$ and $^{17}_{9}F$ form one example (the Z of one equals the N of the other), have ground-state eigenfunctions which are very similar, differing in fact only in respect of the Coulomb interaction. Transitions between the ground states of mirror nuclei should therefore be allowed. This is borne out in the case of the transition

$$^{17}F \rightarrow {}^{17}O + \beta^+.$$

In cases other than mirror nuclei the ground-state eigenfunctions need not have the same degree of similarity, and as a consequence M^*M will be reduced in value. This will lead to higher ft values. In cases where the transition is to an excited state of the daughter nucleus, the eigenfunctions may be markedly different and M^*M very much reduced from unity.

4.15   **Beta-decay selection rules**

If the two states involved in a transition are so different as to have different nuclear spins or different parities, then it follows from the results of quantum mechanics that M, and hence M^*M, are identically zero. This, on the basis of the theory developed above, would lead to an infinite half-life; in other words the transition would be forbidden. Thus the conditions that the transition be allowed are

(a) $\Delta I = 0$, where I is the nuclear spin, and
(b) nuclear parity must not change.

These constitute the *Fermi selection rules*.

It is however found that the violation of these selection rules does not lead to the transition being absolutely forbidden although it is inhibited to the extent of the ft value increasing by about three orders of magnitude. To understand why the ft value does not increase to infinity, we examine the assumption that the β-particle and neutrino wavelengths are of such a magnitude that the eigenfunctions of the outgoing leptons can be assumed constant over nuclear dimensions. This assumption is equivalent to taking only the first term, which is unity, in an expansion of factors of the form $e^{i2\pi r/\lambda}$ describing the spatial shape of the particle waves. If the second term in the expansion be included, then a term has to be added to the right-hand side of equation 4.11 which has a factor

$$\int \psi^*_{Z\pm1,A}\, r\psi_{Z,A}\, d\tau.$$

This integral is not necessarily zero when the violation of the selection rules has resulted in

$$\int \psi_{Z\pm1,A}^{*} \psi_{Z,A}\, d\tau = 0.$$

When the first term, usually the dominant term, is zero and the second term is not zero then the transition is termed *first forbidden*. The magnitude of the *ft* value depends on the magnitude of the second term and is about three orders of magnitude smaller than the *ft* value for allowed transitions. Should the second term as well as the first be zero, then the third term in the expansion has to be taken. It is again about three orders of magnitude smaller than the second term and the transition is now said to be *second forbidden*. This argument can be extended to still higher terms. There are examples (e.g. $^{115}_{49}$In) believed to be as high as fourth forbidden with log *ft* values of about twenty-four.

There is the further complication that there are known cases where the $\Delta I = 0$ and 'no parity change' selection rules are violated yet the *ft* values are as for allowed transitions. To accommodate this fact, it must be assumed that our assumption that the perturbing interaction could be expressed simply as a constant g leading to equation **4.11** is not valid. There may be coupling between the spins of the transforming nucleon and the emitted leptons contributing to the interaction energy. The more complicated matrix elements then arising lead to the *Gamow–Teller selection rules* requiring $\Delta I = 0$ or ± 1 (but *not* $0 \to 0$) with no change of parity. Very many cases of allowed transitions on the basis of *ft* values seem to be governed by the Gamow–Teller rules, although there are a few instances where the Fermi rules are required. For example

$$^{10}C \to {}^{10}B^{*} + \beta^{+} \quad \text{and} \quad {}^{14}O \to {}^{14}N^{*} + \beta^{+}$$

are transitions in which the initial and final states have zero spin yet the *ft* values are as for allowed transitions.

4.16 The theory of electron capture

A theory to describe the process of electron capture can be set up along the same general lines as the theory discussed above for β-decay. There are however two important differences. Firstly, the initial state includes, in addition to the parent nucleus, an orbital electron. This electron will be in a well-defined energy state whose eigenfunction, derived from atomic theory, must be included in ψ_i.

Secondly, there is only one lepton, a neutrino, in the final state. Consequently the density of final states will be given by ρ_f where

$$\rho_f\, dE_f = \frac{L^3 p_\nu^2\, dp_\nu}{2\pi^2\hbar^3}.$$

Thus $\quad \rho_f = \dfrac{L^3 p_\nu^2}{2\pi^2\hbar^3 c} = \dfrac{L^3 E_\nu^2}{2\pi^2\hbar^3 c^3}.$

E_ν has a fixed value, namely the decay energy if the daughter nucleus be formed in the ground state, or the decay energy less the energy of excitation if the daughter nucleus be formed in an excited state.

When allowance is made for these two considerations, it is found that

$$\lambda_{e.c.} = \frac{m_0^5 c^4}{2\pi^3 \hbar^7} g^2 M^* M f_{e.c.},$$

where $\quad f_{e.c.} \simeq 2\pi \left[\frac{\alpha Z}{n} \right]^3 E_\nu^2,$

α being the fine-structure constant, and n the principal quantum number of the shell from which the electron is captured.

A quantity which has been studied extensively by experimental methods is the ratio of electron capture to positron emission in cases where these are competing processes. From the above expression we have in the case of K-capture ($n = 1$)

$$\frac{\lambda_K}{\lambda_{\beta^+}} = \frac{f_K}{f(Z, E_{\beta,\max})} \simeq \frac{2\pi(\alpha Z)^3 E_\nu^3}{f(Z, E_{\beta,\max})},$$

where $f(Z, E_{\beta,\max})$ is given by equation **4.17**.

Note that $E_\nu = E_{\beta,\max} + m_0 c^2$ when the decay energy is great enough for positron emission to be energetically possible.

The theoretical prediction is in reasonably good agreement with the experimental results. K-capture is very much favoured over positron emission for heavy elements, being a thousand times more probably for $Z = 80$ and a decay energy of 1·5 MeV.

We also note that

$$\frac{\lambda_K}{\lambda_L} \simeq 8,$$

since $n = 2$ for the L-shell. This ratio can be altered very much in favour of L-capture when the decay energy is very small. In fact, when the decay energy is so small that atomic binding energies cannot be neglected, L-capture may be energetically possible whereas K-capture is energetically forbidden.

4.17   **Double beta decay**

It was shown in section 4.4 that, having regard to the mass–energy relationships governing β-instability, two neighbouring isobars cannot both be stable. There are however many instances of two stable isobars with an unstable isobar between. For example $^{124}_{50}Sn$ and $^{124}_{52}Te$ are both stable whereas $^{124}_{51}Sb$ is unstable. Now the mass of ^{124}Sn exceeds that of ^{124}Te and energetically the transition ^{124}Sn to $^{124}Te + \beta^- + \beta^-$ is possible. Theoretical estimates have been made of the decay constants for the double β-decay process on the assumption that no antineutrinos are emitted and also on the assumption that two antineutrinos, one associated with each β-particle are emitted. These calculations predicted that in the former case half-lives in the order of 10^{16} years were to be expected, in the latter case half-lives of the order of 10^{22} years.

Attempts have been made, using radiation and particle detectors, to find experimental evidence for this phenomenon, without convincing success. However a lower limit of 10^{16} years has been established for the half-life in several of the isotopes examined. Evidence has also been sought, by analysing with mass spectrometers geological specimens, to establish the extent to which the products of double β-decay may have built up. Such an investigation of tellurium-bearing ores revealed amounts of xenon consistent with a half-life of 10^{21} years for the transition

$$^{130}\text{Te} \rightarrow {}^{130}\text{Xe} + \beta^- + \beta^-.$$

While therefore double β-decay is an important phenomenon in the theory of β-decay, the half-life is known to be so long that for all practical purposes we can consider both nuclei involved to be stable.

4.18 Summary

The theoretical explanation of β-decay transitions introduces the concept of 'weak interaction' between leptons and nucleons which takes its place with gravitation, electromagnetism and the 'strong interaction' between nucleons as one of the four basic interactions in nature. The decay constant for the β-decay process gives information concerning the degree of similarity between initial and final nuclear wave functions. The decay constant for electron capture, equivalent in a sense to positron emission but possible when positron emission is energetically impossible, yields information concerning the wave functions of orbital electrons within the nuclear volume. The β-decay energy enables mass differences to be accurately established when the transition is known to be ground state to ground state. When an excited state is also involved and the β-spectrum is therefore complex, analysis of the β-momentum distribution enables the energy of the excited state to be found.

Chapter 5
Nuclear Mass

5.1 Introduction

In this chapter we discuss nuclear mass, the factors which control its value in the range of nuclei from the very light to the very heavy and its relevance to several nuclear processes. However, as explained in section 3.2, it is the normal practice in nuclear physics to use not the mass of the bare nucleus but rather the mass of the nucleus with its full complement of orbital electrons, i.e. we normally use the mass of the neutral atom.

5.2 The experimental determination of mass values

Experimental information concerning mass values is available from two sources. Firstly it may be derived from the field of mass spectrometry. This technique had its birth in the investigations of canal rays by J. J. Thomson in 1913, and has now been applied to elements throughout the periodic table. The ion of an atom, or more usually of a molecule, is sent through a system of deflecting electric and magnetic fields and from its trajectory its charge-to-mass ratio is measured. Then, providing its charge is known, its mass may be calculated. Usually the mass difference between two ions known to be of almost the same mass value is measured. For example, the molecular ion $^{16}O_2$ and the atomic ion ^{32}S form a *doublet* suitable for a measurement of this kind. A determination of the mass difference then enables the mass of ^{32}S to be accurately related to the mass of ^{16}O.

Secondly, relationships between mass values, of accuracy comparable to that obtained by mass spectrometry, are available from the study of nuclear reactions. For example the measurement of the threshold energy for the photodisintegration of ^2H into a proton and a neutron enables us, invoking the conservation of mass–energy to relate the mass of the deuteron to the masses of the proton and neutron. We recollect that in α- and β-decay a measurement of the particle energies, assuming the distintegration scheme to be known, leads to an estimate of the decay energy from which a relationship between the masses of the parent and daughter nuclei follows.

5.3 Atomic mass unit

For the purpose of discussing absolute masses as distinct from mass differences it is convenient to introduce an *atomic mass unit* rather than to use submultiple units of the gramme or kilogramme. Several such atomic mass units have been proposed and for periods used. The obvious choice of the mass of ^{1}H as one atomic mass unit has the disadvantage that on this scale the mass values of heavy nuclei no longer have the appropriate A-values as the nearest integers. For example, on such a scale ^{208}Pb would have a mass of 206·36 units. This can be avoided by a different choice of unit mass. For many years, the *chemical scale* was based on the natural isotopic mixture of oxygen being sixteen atomic mass units by definition. Alongside this, there previously existed a *physical scale* based on the ^{16}O isotope being defined to be sixteen atomic mass units. Since 1960, an attempt has been made to replace these scales by a new scale based on the isotope ^{12}C being twelve atomic mass units, and this *carbon scale*, or *unified mass scale*, is now customarily used in nuclear physics.

5.4 Binding energy

It is frequently convenient to think in terms of *binding energy* rather than the mass of the nuclear system. The two quantities are of course related by the equation

$$M(Z, A) = ZM_H + NM_n - B(Z, A), \qquad 5.1$$

where $M(Z, A)$ is the mass of the atom whose nucleus contains Z protons and N neutrons, M_H the mass of the hydrogen atom and M_n the mass of the neutron. $B(Z, A)$ is clearly the energy necessary to dissociate the nucleus into its A components. We shall usually express $B(Z, A)$ in units of millions of electronvolts, and we note that the conversion factor to atomic mass units (on the ^{12}C = 12 a.m.u. scale) is 1 MeV = 1·07356 x 10^{-3} a.m.u.

A study of the distribution of the stable nuclei on the nuclear chart described in section 1.4 enables two important deductions to be made about nuclear binding energy. Firstly we note that, considering the light nuclei, the stable isotopes are grouped closely along the line of slope 45°, i.e. they tend to have $Z = N$. We have seen that the stability of these nuclei against β-decay means that their masses are less than the masses of the neighbouring isobars. It follows from equation 5.1 that, if the mass difference exceeds $M_n - M_H$ (i.e. about 0·75 MeV), which is usually the case, then the binding energy of a nucleus with $Z = N$ is greater than the binding energies of the neighbouring isobars. We deduce that equality of proton and neutron numbers enhances binding energy. As we proceed to heavier stable nuclei we notice that N increases more rapidly than Z and must suppose that other factors enter leading to an excess of neutrons. By the time $^{238}_{92}$U is reached we note that this excess is 54, i.e. more than 50 per cent of the total proton number.

Table 1

A even		A odd	
Z even N even	Z odd N odd	Z odd N even	Z even N odd
163	4	49	54

The second important fact emerges from a consideration of the evenness or oddness of the Z- and N-numbers. Table 1 gives the number of stable nuclei (listed in Appendix A) in the four categories arising from the different combinations of evenness and oddness. It is immediately clear that the binding energy in the case of the odd-A nuclei is not affected by whether the odd, or unpaired, nucleon is a proton or a neutron. It is also clear that for even-A nuclei the binding energy is very much affected by the existence of two unpaired nucleons as opposed to the alternative of a complete pairing of nucleons of both kinds. This suggests that there is a pairing energy involved between nucleons of the same kind. We note further that the four exceptions to the general rule that nuclei with Z odd and N odd – which we shall term (odd, odd) nuclei – are not stable are $^2_1$H, $^6_3$Li, $^{10}_5$B and $^{14}_7$N, the four lightest members of the set of nuclei which have both $Z = N$ and Z and N odd. This we can interpret in terms of the increase in the binding energy resulting from the equality of Z and N being more than sufficient to compensate for the loss of binding energy arising from the existence of two unpaired nucleons. For $Z > 7$ this apparently no longer holds.

5.5 Semi-empirical mass formula

The existence of an extensive set of measurements of mass values for both stable and unstable nuclei provides an incentive for the development of a mass formula to fit these experimental data. Weizsacker (1935) employing the analogy between the nucleus and a liquid drop suggested by Bohr, and proceeding on a semi-empirical basis, set up such a mass formula; this, with later modifications, still plays a vital role in systematizing mass values.

Two experimental facts encourage us to make the analogy between the nucleus and a liquid drop. Firstly, as discussed in section 3.14, the radius of a nucleus is with good accuracy found to be given by the formula

$$R = R_0 A^{\frac{1}{3}}.$$

This means that the number of nucleons per unit volume, which equals $A/\frac{4}{3}\pi R^3$, is a constant. The nucleon density thus behaves as does molecule density in a liquid drop, that is, it is independent of the size of the structure. Secondly, as was quickly realized in the early 1930s when the first results of accurate mass spectrometry appeared, the binding energy per nucleon is almost constant over a wide range of nuclei. This property is also shown by molecules in a liquid drop, as is evidenced by latent heat being a general property of the liquid independent of drop size. In both cases the property is interpreted as arising from the forces concerned being short range in nature, resulting in bonds being formed only between close neighbours. The forces are said to *saturate* since, once there is a sufficiency of close neighbours, the binding of one particular component particle is not altered by the existence or absence of more-distant neighbours. Each particle in the assembly thus makes a fixed contribution to the total energy of the system and so the total energy is proportional to the total number of particles in the assembly. If the force did not saturate, then each particle in an assembly of say A particles might be assumed to form bonds with all the other $A - 1$ particles in the assembly. The total number of bonds formed would then be $\frac{1}{2}A(A - 1)$. Assuming a certain fixed energy to be attributable to each bond, the total energy would then be proportional to $A(A - 1)$. Consequently the binding energy per particle would be proportional to $A - 1$, that is, it would be expected to increase with A. This is at variance with the facts, which favour the saturation hypothesis for nuclear matter.

We therefore begin the construction of the mass formula by taking a main term in the binding energy proportional to A. This term we represent by $a_v A$, where a_v is a constant, and we refer to $a_v A$ as the *volume energy*. Now in any nucleus of a finite size some of the nucleons must lie on the surface and have a different arrangement of closest neighbours from those nucleons which lie in the interior. The same situation arises of course in the case of a liquid drop, where the fact that the molecules on the surface are differently arranged, with respect to nearest neighbours, from those in the volume of the drop gives rise to the phenomenon of surface tension. Under the action of surface tension drops take up a shape which minimizes the surface area and maximizes the total binding energy. In the nuclear case we thus have to correct the binding energy (in the belief that a nucleon on the surface will make a smaller contribution to the total binding energy) by an amount proportional to the surface area. As the radius is proportional to $A^{\frac{1}{3}}$, the surface area can be taken to be proportional to $A^{\frac{2}{3}}$ and the *surface energy* contribution we write as $- a_s A^{\frac{2}{3}}$.

So far it has been assumed that the nuclear system is held together by an attractive force between nucleons which acts irrespective of their identity as protons or neutrons. In addition to this cohesive nuclear force, there will be the Coulomb force acting between protons. This is a long-range repulsive force and hence reduces the total binding energy. The term to represent this effect, *the Coulomb energy*, can be calculated from the principles of elementary electrostatics if the spatial arrangement of protons in the nucleus is known. If we assume the protons to be uniformly distributed throughout the nuclear volume, we may then imagine the protons in the nucleus to be assembled in spherical layers. Assume that at an intermediate stage of formation the nucleus has radius r and a layer of thickness dr is brought up, proton by proton, from infinity. Let ρ_p be the number of protons per unit volume. Then the charge already in the partially assembled nucleus will be $\frac{4}{3}\pi r^3 \rho_p e$. The work done against the Coulomb force in bringing one additional proton from infinity will then be

$$\int_\infty^r \frac{4\pi r^3 \rho_p e^2}{3x^2} \, dx = \tfrac{4}{3} \pi r^2 e^2 \rho_p$$

in magnitude. In this layer there will be $4\pi r^2 \, dr \, \rho_p$ protons. Thus the energy built into this layer is

$$\tfrac{16}{3} \pi^2 r^4 \rho_p^2 e^2 \, dr.$$

We therefore see that the Coulomb energy built into the nucleus when it is assembled to a radius R will be

$$\int_0^R \frac{16}{3} \pi^2 r^4 \rho_p^2 e^2 \, dr = \frac{16}{15} \pi^2 \rho_p^2 R^5 e^2.$$

Now $\quad \rho_p = \dfrac{Z}{\frac{4}{3} \pi R^3}.$

Substituting this value into the above equation we have therefore the result

$$\text{Coulomb energy} = \frac{3}{5} \frac{Z^2 e^2}{R}$$

$$= \frac{3}{5} \frac{Z^2 e^2}{R_0 A^{\frac{1}{3}}},$$

for the uniform distribution assumed.

If the protons are not uniformly distributed, then the form of dependence on Z and A is the same but the numerical coefficient is different. For example, if all the protons were on the surface, the Coulomb energy would be that for a charge Ze on a conducting sphere of radius R. The sphere would have an electrical capacity equal to R and hence its energy when a charge Ze is placed on it is from elementary electrostatics given by $\frac{1}{2} Z^2 e^2 / R$. We therefore introduce in the general case a term, again negative since it represents a disruptive effect, into the binding energy equal to $- a_C Z^2 / A^{\frac{1}{3}}$ to represent the Coulomb energy. In deriving this expression, it has been assumed that even within one proton there is a certain Coulomb energy associated with one 'part' of the basic charge interacting with another. Whether there is such a contribution to the energy or not is a basic assumption to be built into any nuclear model. If there is no such internal Coulomb energy to be associated with the proton, then from the above expression we must subtract a contribution $- a_C / A^{\frac{1}{3}}$, (obtained by putting $Z = 1$) for each proton in the assembly. On this assumption therefore, the Coulomb energy would be $- a_C Z(Z - 1)/A^{\frac{1}{3}}$. The difference between this and the original expression becomes less important as Z increases and in what follows we retain the original expression $- a_C Z^2 / A^{\frac{1}{3}}$ for the Coulomb energy.

So far we have considered the nucleus in terms of classical physics and on that basis there is no explanation for the equality of Z and N leading to particularly stable nuclear configurations. Rather, in view of the Coulomb energy, an excess of neutrons should result even in the case of light nuclei. The fact that light nuclei do not show neutron excess leads us to introduce into the binding energy, on an empirical basis, *an asymmetry energy* which is negative for $Z \neq N$ and zero for $Z = N$. A rough justification for this and an indication of the possible form of this term may be given by the following argument. The nucleons, which we assume to be obeying the laws of quantum mechanics, must be occupying states of definite energy. In terms of the Pauli exclusion principle, only one nucleon of one kind can occupy one state. The lowest energy states will be filled first. In so far as we can neglect Coulomb effects, we may take the energy states to be similar for neutrons and protons. To go on adding particles of one kind in constructing heavier nuclei thus involves filling higher energy states appropriate to that particle while lower energy states appropriate to particles of the other kind remain vacant. Thus, if there is a neutron excess of $N - Z$, which we assume even, it means that there are $N - Z$ neutron states filled above the last filled proton state. If now the neutrons in the top half of this range of states, i.e. the top $\frac{1}{2}(N - Z)$ neutrons were to be transformed into protons then each could drop down $\frac{1}{2}(N - Z)$ states. If the states were evenly spaced and energy ϵ apart, the energy gained per nucleon would be $\frac{1}{2}(N - Z)\epsilon$ and the total energy gained would thus be $\frac{1}{4}(N - Z)^2 \epsilon$. This substitution of protons for neutrons of course has the effect of increasing the Coulomb-energy term, and the N, Z values which lead to maximum binding energy will be determined by minimizing the net result of the opposed effects of asymmetry and Coulomb energy. It is believed that the average energy spacing ϵ for the last few nucleons is approximately proportional to $1/A$ and hence the asymmetry energy term is substituted into the binding energy in the form $-a_a(N - Z)^2/A$.

Finally we have to have regard to the stability of (even, even) as compared to (odd, odd) nuclei noted in section 5.4. The pairing energy which was suggested by the information in Table 1 is allowed for by introducing into the formula for the binding energy a term δ, which is taken positive for (even, even) nuclei, zero for (odd, even) and (even, odd), and negative for (odd, odd) nuclei. Pairing energy is a concept which is added for purely empirical reasons to the liquid-drop model and hence the model cannot pronounce on the form δ should take. Other models, for example the shell model, have been appealed to and various expressions for δ have been suggested. The form

$$\delta = a_P \frac{1}{A^{\frac{3}{4}}}$$

has been commonly used and we shall insert that form in what follows. Collecting together the various terms introduced above and substituting in equation 5.1 we have for the mass of a nucleus charge number Z, mass number A

$$M(Z, A) = ZM_H + NM_n - a_v A + a_s A^{\frac{2}{3}} + a_C \frac{Z^2}{A^{\frac{1}{3}}} + a_a \frac{(N - Z)^2}{A} \pm \delta$$

$$= AM_n + Z(M_H - M_n) - a_v A + a_s A^{\frac{2}{3}} + a_C \frac{Z^2}{A^{\frac{1}{3}}} + a_a \frac{(A - 2Z)^2}{A} \pm \delta.$$

$$5.2$$

We note that in this expression there are five adjustable constants. In principle these can be found from five simultaneous equations formed by substituting five known mass values. The usefulness of the formula and the validity of the physical arguments employed in its construction are then to be judged by how well it predicts the hundreds of other mass values which have been measured and by how well it predicts the mass differences involved, and hence energy released, in very many nuclear reactions. Various sets of values for the constants have been suggested, differing slightly depending on the range of nuclei which were under investigation. We shall follow R. D. Evans (*The Atomic Nucleus*, McGraw-Hill, 1955) in taking for the constants the following set of values which result in reasonably good agreement with measured mass values over the whole range of A-values:

$$a_v = 14.1 \text{ MeV}; \quad a_s = 13 \text{ MeV}; \quad a_C = 0.595 \text{ MeV};$$

$$a_a = 19 \text{ MeV}; \quad \delta = \frac{33.5}{A^{\frac{3}{4}}} \text{ MeV}.$$

5.6   Binding energy per nucleon

An impression of the relative importance of the contributions to the binding energy of a nucleus made by the various terms in the mass formula, and the change in their relative importance as we proceed from light to heavy nuclei, is obtained from Table 2 and from Figure 14. As the surface-energy contribution falls, it is largely compensated by increased Coulomb energy. The important observed fact that middle-weight nuclei have a slightly greater binding energy per nucleon than either heavier or lighter nuclei is reproduced by the formula. Arising from the interplay of the surface and Coulomb terms, it is seen that energy may be released by fusing lighter nuclei or by dividing heavier nuclei. The effect of the Coulomb barrier however inhibits these processes, otherwise all material would tend to transform so as to end in middle-weight elements.

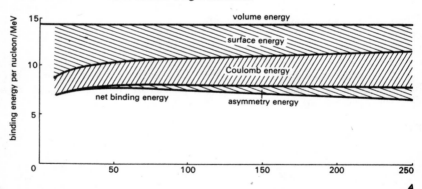

Figure 14 The different negative contributions to the binding energy per nucleon successively subtracted from a constant-volume energy per nucleon to leave the net binding energy per nucleon, all plotted as a function of the mass number A

Table 2

	Z	A	Volume B_v/A $a_v = 14.1$ MeV	Surface B_s/A $a_s = 13$ MeV	Coulomb B_C/A $a_C = 0.595$ MeV	Asymmetry B_a/A $a_a = 19$ MeV	B/A (calculated) /MeV	B/A (experimental) /MeV
O	8	17	14.1	5.06	0.87	0.07	8.10	7.75
S	16	33	14.1	4.05	1.44	0.02	8.59	8.50
Mn	25	55	14.1	3.42	1.78	0.16	8.74	8.75
Cu	29	65	14.1	3.23	1.91	0.22	8.74	8.75
I	53	127	14.1	2.59	2.62	0.52	8.37	8.43
Pt	78	195	14.1	2.24	3.20	0.76	7.90	7.92
Bk	97	245	14.1	2.08	3.65	0.82	7.55	7.52

5.7 Mass surface

Since $A = Z + N$, we can regard equation 5.2 as expressing the mass of a nucleus in terms of the two parameters (Z, N). If now, on the plot of the nuclei which has N along the x-axis and Z along the y-axis, we imagine verticals erected along the z-axis of length proportional to $M(Z, N)$, then the end points of these verticals will define a surface. This we refer to as the *mass surface*.

5.8 Mass excess

The masses, as was noted in section 5.3, are in all cases quite close to integral values when expressed in atomic mass units. However, in nuclear radioactive transformations and in nuclear reactions it is mass differences we are concerned with, and the important information then lies entirely in the amounts by which the masses depart from integral values. It is consequently convenient to work with the quantity $M(Z, N) - A$, where the mass is in atomic mass units and this quantity is referred to as the *mass excess*. By virtue of the relationship $E = mc^2$ between energy and mass, the quantity can also be expressed in energy units.

It may easily be confirmed that in equation 5.2 we may substitute on the left-hand side the mass excess for $M(Z, A)$ providing we replace the mass M_H by the mass excess of the hydrogen atom and M_n by the mass excess of the neutron.

5.9 Mass parabolas

We now consider the relationship that equation 5.2 predicts between the masses of isobars. To do so, we eliminate N through the relationship $N = A - Z$ and regard, for a particular set of isobars, A as a constant. We then have that

$$M(Z, A) = B + CZ + DZ^2,$$ 5.3

where $B = AM_n - a_v A + a_s A^{\frac{2}{3}} + a_a A \pm a_P \dfrac{1}{A^{\frac{3}{4}}},$

$C = M_H - M_n - 4a_a,$

$D = \dfrac{a_C}{A^{\frac{1}{3}}} + \dfrac{4a_a}{A}.$

The section of the mass surface taken through isobars is therefore seen to be parabolic in shape.

First we consider isobars corresponding to odd values of A. For these $a_P = 0$. The coefficients in equation 5.3 are then single valued and the nuclear-mass values lie on one parabola. Having regard to the conditions for β-decay, we see that there will be only one stable nucleus in this set; it will be the nucleus with the smallest mass value and therefore the nucleus whose mass value is closest to the vertex of the parabola. All other members of the set have mass relationships with respect to a neighbour which permit β-decay (or electron capture) to that neighbour. The vertex of the parabola will be given by $Z = Z_A$, say, where

$$\left[\frac{\partial M}{\partial Z}\right]_{Z=Z_A} = C + 2DZ_A = 0.$$

i.e. $\qquad Z_A = -\dfrac{C}{2D} = \dfrac{(M_n - M_H) + 4a_a}{2(a_C/A^{\frac{1}{3}} + 4a_a/A)}.$

From the mass tables we have $M_n - M_H = 0.7824$ MeV. Using this, together with the values of a_a and a_C quoted above, we have

$$Z_A = \frac{76.7824}{2(0.595/A^{\frac{1}{3}} + 76/A)}. \qquad\qquad 5.4$$

We now take as an example the case of isobars with $A = 141$. For this value of A, equation 5.4 gives Z_A as 58.76. From the table of isotopes it will be found that the stable member of this set of isobars is $^{141}_{59}$Pr, which corresponds to the mass value closest to the vertex of the parabola. In Figure 15 the experimental masses of the other members of this isobaric set are plotted together with the section of the mass surface.

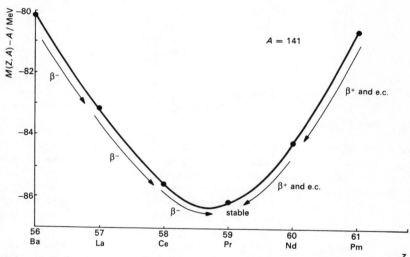

Figure 15 Mass excess of a set of odd-A isobars defining a single sheet of the mass surface when plotted against the charge number Z

Turning to isobars with even A-values, we no longer have $a_P = 0$. In equation 5.3 B has then two values, giving rise to two parabolas, an upper parabola corresponding to (odd, odd) nuclei and a lower corresponding to (even, even) nuclei. Note that the vertices of the two parabolas have the same value of Z. The (even, even) nucleus whose mass number is closest to Z_A should be stable. We take as an example the case of $A = 134$ giving $Z_A = 56.17$. $^{134}_{56}$Ba is in fact found to be stable in accordance with this prediction. However, a second member of the

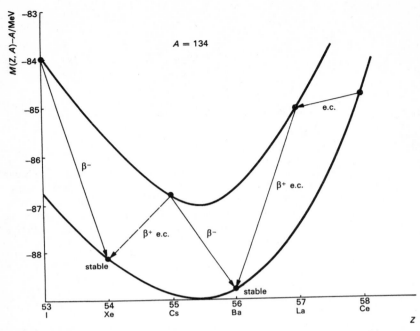

Figure 16 Mass excess of a set of even-A isobars. The even-Z, even-N nuclides define one sheet of the mass surface, the odd-Z, odd-N nuclides define a separate sheet lying above the first

set, $^{134}_{54}$Xe, is also stable. The experimental mass values for these isobars are plotted in Figure 16 together with the two sections of the mass surface. The reason for the stability of $^{134}_{54}$Xe, which lies on the lower parabola, is there seen to be due to the position of its neighbours which lie on the upper parabola. Decay of this isotope to $^{134}_{56}$Ba would only be possible by double β-decay which, it was noted in section 4.17, is likely to be unobservable by virtue of the very long half-life involved. It is interesting to note that $^{134}_{55}$Cs satisfies the conditions for decay both by β⁻ emission and by β⁺ emission. In the case of this nucleus, β⁻ emission is so favoured by the selection rules that it alone is observed. There are however other examples of nuclei similarly situated with respect to neighbouring isobars which do exhibit β⁻ emission as well as β⁺ emission and electron capture. $^{64}_{29}$Cu is a well-known example of such a nucleus.

5.10 Stability against alpha decay

The mass formula may be used to decide whether or not a nucleus is stable against transformation by α-particle emission. It follows from the discussion of section 3.2 that the stability is determined by the algebraic sign of

$$Q = M_{\text{parent}} - M_{\text{daughter}} - M_{\text{He}}$$
$$= M(Z, A) - M(Z - 2, A - 4) - M(2, 4).$$

Treating Z and A as continuous variables, we have

$$M(Z, A) - M(Z - 2, A - 4) = \frac{\partial M}{\partial Z} \Delta Z + \frac{\partial M}{\partial A} \Delta A, \qquad 5.5$$

where on the right-hand side Z and A are to be averaged as between the parent and daughter values.

Thus, from equations 5.2 and 5.5, and taking nuclei with odd A-values so that the pairing energy δ may be ignored, we have

$$Q = \frac{4a_C Z}{A^{\frac{1}{3}}} - 4a_v + \frac{8}{3} a_s A^{-\frac{1}{3}} - \frac{4}{3} a_C \frac{Z^2}{A^{\frac{4}{3}}} - 4a_a \frac{(A - 2Z)^2}{A^2} + B(^4\text{He}).$$

Substituting the values quoted above for the constants and introducing the experimental value for the binding energy of the α-particle, we have

$$Q = \frac{Z}{A^{\frac{1}{3}}} \left(2 \cdot 38 - 0 \cdot 793 \frac{Z}{A} \right) + \frac{34 \cdot 67}{A^{\frac{1}{3}}} - 76 \left(1 - \frac{2Z}{A} \right)^2 - 28 \cdot 1.$$

If into this expression we substitute the Z- and A-values for $^{141}_{59}\text{Pr}$, then

$$Q = -0 \cdot 24 \text{ MeV}.$$

If the values for $^{145}_{60}\text{Nd}$ are used, then $Q = +0 \cdot 92$ MeV.

For $^{191}_{77}\text{Ir}$, $Q = +2 \cdot 62$ MeV.

Thus the formula with the values of the constants chosen above predicts that for A-values in excess of about 145, α-decay is becoming energetically possible. Note that the term making the largest contribution to the change in the Q-value as we proceed to heavier nuclei is that associated with the Coulomb energy, that is, it is the electrostatic repulsion between protons that is leading to α-instability. It does not follow that if α-decay is energetically possible it will be experimentally observed. With very low Q-values, the outgoing α-particle will require to penetrate a very wide barrier and, as was discussed at length in section 3.13, this will give rise to very long half-lives. We note that this general prediction concerning the onset of α-decay as we proceed up the periodic table is in line with the observed distribution of α-emitters plotted in Figure 5.

5.11 Stability of nuclei against spontaneous symmetric fission

As a further example of the application of the mass formula, we consider the stability of a nucleus against undergoing *spontaneous symmetric fission*. This process involves a nucleus (Z, A) splitting into two identical nuclei, called *fission fragments*, each having charge number $\frac{1}{2}Z$ and mass number $\frac{1}{2}A$, Z and A both being assumed even. The Q-value for such a transformation will be given by

$$Q = M(Z, A) - 2M(\tfrac{1}{2}Z, \tfrac{1}{2}A),$$

and the transformation will be energetically possible if $Q > 0$.

Now, substituting the appropriate values of Z and A into equation 5.2 leads to

$$Q = a_s A^{\frac{2}{3}}(1 - 2^{\frac{1}{3}}) - a_C \frac{Z^2}{A^{\frac{1}{3}}}(1 - 2^{-\frac{2}{3}})$$

$$= -3\cdot38\ A^{\frac{2}{3}} + 0\cdot22\ \frac{Z^2}{A^{\frac{1}{3}}}.$$ 5.6

Q will thus be positive for nuclei satisfying

$$\frac{Z^2}{A} > \frac{3\cdot38}{0\cdot22} = 15\cdot36.$$

This inequality is satisfied for $Z = 35, A = 79$ (i.e. ^{79}Br) and for heavier nuclei. Once again we note that the process being energetically possible does not mean that it is observed. The fission fragments are very highly charged and we must have regard to the Coulomb barrier that will exist. To see how to make allowance for this imagine the process to be reversed, i.e. imagine the fission fragments, each having a charge $\frac{1}{2}Ze$ and being spherical with a radius $R_0(\frac{1}{2}A)^{\frac{1}{3}}$, brought towards each other from infinite separation. As they approach, the energy of the system is increased by virtue of the work done against the Coulomb repulsion. If we ignore any deformation of shape of the fission fragments and assume that the only forces involved until the spherical nuclei touch are the Coulomb forces, then the work done against the Coulomb forces in bringing the fission fragments from infinite separation until they are in contact will be

$$\frac{(\frac{1}{2}Ze)^2}{2R_0(\frac{1}{2}A)^{\frac{1}{3}}} = 0\cdot1532 \times \frac{Z^2}{A^{\frac{1}{3}}}\ \text{MeV}.$$

It is assumed that the short-range nuclear forces take over when the distance between centres is less than twice the fission-fragment radius and that the potential energy will drop as the fragments coalesce. The value to which the potential energy drops will be the fission energy given by Q in equation 5.6. Now, if the potential barrier is not penetrable and the fission is to be truly spontaneous, the peak potential energy reached, namely that at separation $2R_0(\frac{1}{2}A)^{\frac{1}{3}}$, should equal the fission energy, i.e. there should be no drop in potential energy as the fragments coalesce. Thus Q should be greater than $0\cdot1532 \times Z^2/A^{\frac{1}{3}}$ MeV and not simply greater than zero. Incorporating this more stringent condition, then from equation 5.6 we have

$$\frac{Z^2}{A} > \frac{3\cdot38}{0\cdot0668} = 50\cdot6.$$

For the lead isotopes Z^2/A is approximately 32, for uranium it is 36 and for lawrencium it is 41. Thus the prediction of the mass formula, with the choice of values of constants made above, is that spontaneous symmetric fission will not take place for nuclei with a Z-value less than about 110. It is to be noted however that there is the possibility of spontaneous *asymmetric* fission and the possibility

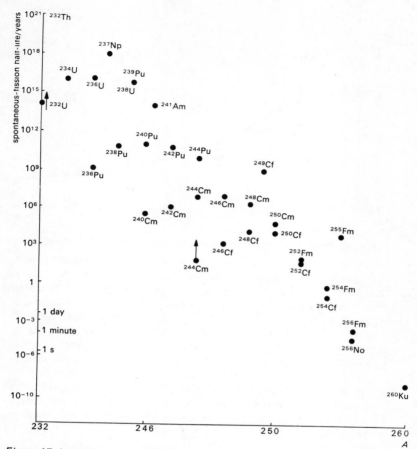

Figure 17 Logarithmic plot of spontaneous fission half-life against mass number, based on data of Figure 86

of barrier penetration to be taken into account. We shall not attempt to treat these in terms of the simple model here developed, but note that spontaneous fission is observed for Z-values of 92 and upwards. In Figure 17 the half-life against spontaneous fission is plotted as a function of A. It is clear that the probability of spontaneous fission is one of the factors setting a limit to an extension of the observed heavy nuclei to even higher A-values.

5.12 Induced fission

When the potential energy of the fission fragments is considered, as it was above, as a function of their separation, the energy at zero separation, while below the

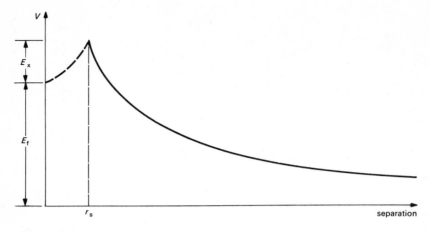

Figure 18 Potential energy as a function of separation of fission fragments

peak which occurs at a separation of about twice the fission-fragment radius, may still be very far above the energy corresponding to infinite separation. This is illustrated in Figure 18. If the original nucleus is disturbed so that the fragments only separate within certain limits ($r < r_s$) then the nucleus will not undergo fission. If the disturbance is such as to separate the fragments instantaneously by a distance greater than r_s, then fission will result with the fragments going off to infinity. The energy released in that case will be E_f. Thus a trigger energy or excitation energy of E_x can release the fission energy of E_f. For the heaviest nuclei this situation can arise by the capture of a neutron of effectively zero energy. When the neutron is captured the neutron binding energy, which as we have seen is a little less than 8 MeV, then provides this excitation and the full fission energy of about 200 MeV may then be released mainly as kinetic energy of the fission fragments.

5.13 Summary

By appeal to a simple classical model with an empirical overlay of quantum-mechanical effects, the semi-empirical formula arrived at permits, in terms of five adjustable parameters, a description to be given of a mass surface which, with exceptions to be noted in the next chapter, gives a satisfactory fit to the experimental values for many hundreds of nuclei. Using the formula, predictions can be made concerning the stable members to be expected in a set of isobaric nuclei. Criterion of stability against α-decay and spontaneous fission can be arrived at which lead to an explanation of why these processes are limited to particular ranges of A-values. Also some insight is given into the balance of the different contributions to binding energy and into the change in this balance as one proceeds from light to heavy nuclei.

Chapter 6
Nuclear Shell Model

6.1 Introduction

In Chapter 5 the liquid-drop model was developed as a basis for the discussion of a number of nuclear properties, in particular binding energy. This model will again be used in a later chapter to explain further nuclear properties, for example nuclear fission. However, there are certain properties, one of which is the important property of angular momentum, which cannot find a place in any elementary way in the framework of the liquid-drop model. We now proceed to outline a model which developed in parallel with the liquid-drop model and which plays a very important role in certain areas of nuclear physics. We shall see that it depends on assumptions which appear incompatible with those of the liquid-drop model. The reason for these two models, based on apparently contradictory assumptions, each having its areas of useful application, has for long been a central problem in nuclear physics and is a topic to which we return in a later chapter.

6.2 Experimental evidence for 'magic' numbers

Evidence from several different fields of study can be assembled to show that certain values of Z and N, the proton and neutron numbers of a nucleus, confer special properties. These Z- and N-values, which are referred to as the *'magic' numbers*, are 2, 8, 20, 28, 50, 82 and 126. We now collect some of the more important strands of the evidence for the existence of these magic numbers.

When the adjustable constants in the semi-empirical mass formula of section 5.5 are chosen for the best general fit to experimentally measured mass values, it is found that the greatest discrepancies are in regions corresponding to magic numbers of protons or neutrons. Whereas the formula reproduces the general trend of the mass surface to an accuracy of 1 or 2 MeV, in the neighbourhood of magic numbers the experimental mass values fall about 10 MeV below the mass-formula values. Thus the indications are that a nucleus with a magic number of neutrons or protons has an unusually large binding energy.

This high binding energy brings in its train several other effects. For example, an examination of the nuclear chart shows that the element with the largest number of isotopes is tin, for which $Z = 50$, while the neutron number corresponding to the greatest number of isotones is $N = 82$.

Estimates of the relative abundance of elements in the universe, based on the chemical analysis of meteorites reaching the earth, on the spectral analysis of solar and stellar bodies, on the spectrum of nuclei in primary cosmic radiation and on studies of the overall composition of the Earth's crust, results in a plot of relative atomic abundance against mass number A, which shows peaks where 50, 82 or 126 nucleons are involved (Alpher and Herman, 1953). There is a very marked peak corresponding to ^{56}Fe which, it is interesting to note, is the end product of the decay of $^{56}_{28}$Ni which has $Z = N = 28$. Nuclei which, like $^{56}_{28}$Ni, have a magic number of both neutrons and protons are referred to as *doubly magic*.

Another pointer comes from the field of natural radioactivity. There exist three naturally occurring radioactive series, the thorium series based on the long-lived parent $^{232}_{90}$Th, the uranium series based on $^{238}_{92}$U and the actinium series based on $^{235}_{92}$U. These series terminate in $^{208}_{82}$Pb, $^{206}_{82}$Pb and $^{207}_{82}$Pb respectively, all these terminal nuclides having 82 protons. ^{208}Pb has 126 neutrons in addition, and is therefore doubly magic.

Particularly convincing evidence for 'magic' properties comes from the capture probability for slow neutrons. This is a relatively unlikely process for nuclei having a magic number of neutrons, whereas for nuclei having one neutron less than a magic number it is a highly likely process. For example, neutron capture into the nucleus $^{135}_{54}$Xe having 81 neutrons is seven orders of magnitude more likely than capture into $^{136}_{54}$Xe, which has 82 neutrons. The behaviour of $^{135}_{54}$Xe is of great practical importance. It is a fission product of uranium and, with its great appetite for slow neutrons, constitutes a serious source of 'poisoning' in nuclear reactors as it accumulates with operating time.

Nuclei having one neutron more than a magic number also have distinctive properties. They exhibit delayed neutron emission. $^{137}_{54}$Xe is one example of this. When it is formed by the β-decay of $^{137}_{53}$I it is frequently in an excited state with an energy of excitation higher than the energy of attachment of the least tightly bound neutron. As a result, rather than the excited state decaying to the ground state with the emission of a photon, neutron emission takes place from the excited state. These neutrons are delayed because of their association with the β-decay. ^{137}I is produced in a nuclear reactor as one of the many fission products. It has a half-life of twenty-four seconds. Its presence in the reactor means that when the chain reaction is stopped by, say, the insertion of control rods, the neutron population does not promptly drop to zero. A certain component of this population associated with delayed neutron emitters like ^{137}I falls off with the half-life of these isotopes. This has an important bearing on the problem of adjusting reactor operating levels as it affects the speed of response of the neutron population to the control settings. Another delayed neutron emitter is $^{87}_{36}$Kr. In this case the phenomenon is associated with the low value of neutron binding in the case of the fifty-first neutron.

The same effects are exhibited if one considers directly the neutron attachment energy. In Figure 19 this quantity is shown on a section of the nuclear chart. It will be observed, by considering the neutron separation energy for each of the chemical elements in the plot, that there is in each case a sharp drop in energy

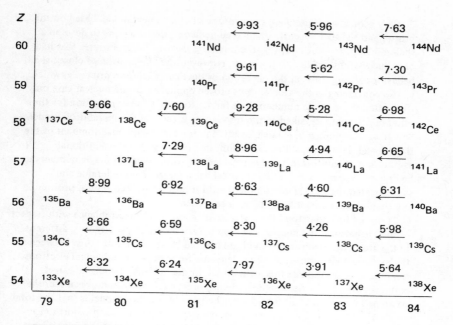

Figure 19 Neutron binding energy is entered to the left of each nuclide to show discontinuity at the magic number, 82, of neutrons in the nucleus

when the eighty-third neutron is reached.

The effects listed to this point all stem from the increased binding energy which a magic number of nucleons confers. There are other effects to be noted which are associated with angular momentum. The pattern of 'spin' values (i.e. intrinsic angular momentum) of the ground states of stable nuclei changes as magic numbers of nucleons are reached. We consider this in detail in discussing the applications of the shell model. It is also to be noted that there are 'islands of isomerism' in areas of the nuclear chart related also to magic numbers of nucleons. The existence of these islands depends on the spin of excited states of nuclei being very different from the ground-state spins and this phenomenon also will be discussed in detail later.

6.3 Nuclear shells and their atomic analogue

Nuclear behaviour with respect to magic numbers of nucleons is reminiscent of the behaviour of atoms with respect to closed shells of electrons. For example, the behaviour of the neutron separation energy as one goes from ^{139}Ce through ^{140}Ce to ^{141}Ce recalls the behaviour of the first ionization energy, which is the measure of the electron separation energy, as one goes from chlorine through argon to potassium. It has become customary to refer to a magic number of nucleons as a closed shell of nucleons in analogy with the electron shells in an atom.

Very soon after Heisenberg's first proposal of a nuclear model based on the proton and the then recently discovered neutron came attempts to develop a shell model of the nucleus using the quantum-mechanical procedures that had been successfully applied to the atom. However, the explanation of closed-shell behaviour arising through the operation of the Pauli exclusion principle, which was so convincing in the atomic case, proved applicable in the nuclear case only to the first three magic numbers. The failure to provide an explanation for the other magic numbers, allied to the resounding success of the liquid-drop model which was developing in parallel, resulted in the temporary abandonment of the shell model. It was revived in 1945 with the discovery that the additional assumption of *spin–orbit coupling* enabled the whole range of magic numbers to be derived. Subsequently the shell model has undergone considerable and sophisticated theoretical development until it now occupies a central position in any theoretical discussion of the nuclear system.

It should be understood that, despite their superficial resemblance with respect to shell behaviour, the nuclear and atomic systems are physically very different. In the atom, the electron motion is dominated by the Coulomb force between the individual electrons and the nucleus; the force between individual electrons is a small perturbation of this main effect. In the nucleus, there is no effect corresponding to the dominant Coulomb force. Each nucleon moves under the combined influence of all the others. The basis of the shell model is that the total effect of all the other nucleons can be represented, in so far as the short-range nuclear interaction is concerned, by a smoothly varying potential having a large negative value in the central region of the nucleus and rising to zero at the nuclear surface. The general features of the shell model should then emerge from a consideration of the motion of a nucleon in this averaged potential.

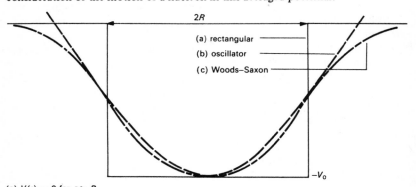

(a) rectangular
(b) oscillator
(c) Woods–Saxon

(a) $V(r) = 0$ for $r \geqslant R$
   $V(r) = -V_0$ for $r < R$

(b) $V(r) = V_0 \left[\frac{1}{2}\left(\frac{r}{R}\right)^2 - 1 \right]$

(c) $V(r) = \frac{-V_0}{1 + e^{(r-R)/a}}$ with $a = \frac{R}{5}$

Figure 20 Three different shapes of potential well used in nuclear theory

Various shapes for the nuclear potential considered as a function of distance from the nuclear centre have been suggested and used. We begin by considering the potential to have the shape illustrated in Figure 20(a), that is, constant throughout the nuclear volume rising infinitely steeply to zero at the nuclear surface. The motion of a nucleon in this spherical rectangular potential well will now be investigated.

6.4   The theory of the spherical rectangular potential well

Let the potential well be represented by $V(r)$, where $V(r) = 0$ for $r \geqslant R$, and $V(r) = -V_0$, where V_0 is a positive constant, for $r < R$. This well is represented in Figure 20. The wave function of a nucleon in a stationary state in the well must be a solution of the time-independent Schrödinger equation

$$\left[\frac{\partial^2}{\partial x^2} + \frac{\partial^2}{\partial y^2} + \frac{\partial^2}{\partial z^2}\right]\psi + \frac{2M}{\hbar^2}[W - V(r)]\psi = 0, \qquad \qquad 6.1$$

where M is the nucleon mass and W its total energy. W must have a value such that ψ is zero at the nuclear surface.† ψ is then termed an *eigenfunction*, and W the corresponding *eigenvalue*. Because of the spherical symmetry of the problem, it is convenient to transform to the usual spherical coordinates using the transformations

$$x = r\sin\theta\cos\phi, \qquad y = r\sin\theta\sin\phi, \qquad z = r\cos\theta.$$

When we change the variables in equation 6.1, it becomes

$$\left[\frac{\partial^2}{\partial r^2} + \frac{2}{r}\frac{\partial}{\partial r} + \frac{1}{r^2}\frac{\partial^2}{\partial\theta^2} + \frac{\cot\theta}{r^2}\frac{\partial}{\partial\theta} + \frac{1}{r^2\sin^2\theta}\frac{\partial^2}{\partial\phi^2}\right]\psi + \frac{2M}{\hbar^2}[W - V(r)]\psi = 0.$$
$$6.2$$

If the potential is a function of r only, i.e. $V(r)$ does not depend on θ or ϕ, then a separation of the three variables can be achieved. This we develop in two stages. Firstly let

$$\psi(r, \theta, \phi) = R(r)\,Y(\theta, \phi).$$

Then equation 6.2 may be written

$$Y\frac{d^2R}{dr^2} + \frac{2Y}{r}\frac{dR}{dr} + \frac{R}{r^2}\frac{\partial^2Y}{\partial\theta^2} + \frac{\cot\theta}{r^2}R\frac{\partial Y}{\partial\theta} + \frac{R}{r^2\sin^2\theta}\frac{\partial^2Y}{\partial\phi^2} +$$
$$+ \frac{2M}{\hbar^2}[W - V(r)]\,RY = 0.$$

Multiplying by r^2/RY and rearranging the terms we have

$$\frac{r^2}{R}\frac{d^2R}{dr^2} + \frac{2r}{R}\frac{dR}{dr} + \frac{2Mr^2}{\hbar^2}[W - V(r)] = -\frac{1}{Y}\left[\frac{\partial^2Y}{\partial\theta^2} + \cot\theta\frac{\partial Y}{\partial\theta} + \frac{1}{\sin^2\theta}\frac{\partial^2Y}{\partial\phi^2}\right].$$

† This is strictly true only in the limit $V_0 \rightarrow \infty$.

Now the right-hand side depends on θ and ϕ but is independent of r, while the left-hand side depends only on r. It follows that each side must equal a constant independent of r, θ and ϕ. Let this constant be $l(l + 1)$ and we then have the two equations

$$\frac{r^2}{R} \frac{d^2 R}{dr^2} + \frac{2r}{R} \frac{dR}{dr} + \frac{2Mr^2}{\hbar^2} [W - V(r)] = l(l + 1) \qquad 6.3$$

and $\quad \dfrac{1}{Y} \left[\dfrac{\partial^2 Y}{\partial \theta^2} + \cot \theta \, \dfrac{\partial Y}{\partial \theta} + \dfrac{1}{\sin^2 \theta} \, \dfrac{\partial^2 Y}{\partial \Phi^2} \right] + l(l + 1) = 0 \qquad 6.4$

If it now be assumed that the function Y can be expressed as the product of two functions, one, Θ, a function of θ alone, and a second, Φ, a function of ϕ alone, we can express equation 6.4 as

$$\frac{1}{\Theta \Phi} \left[\Phi \frac{d^2 \Theta}{d\theta^2} + \Phi \cot \theta \frac{d\Theta}{d\theta} + \frac{\Theta}{\sin^2 \theta} \frac{d^2 \Phi}{d\phi^2} \right] + l(l + 1) = 0.$$

Multiplying by $\sin^2 \theta$ and rearranging the terms this becomes

$$\sin^2 \theta \left[\frac{1}{\Theta} \frac{d^2 \Theta}{d\theta^2} + \frac{1}{\Theta} \cot \theta \frac{d\Theta}{d\theta} + l(l + 1) \right] = -\frac{1}{\Phi} \frac{d^2 \Phi}{d\phi^2}.$$

Each side is seen to be a function of one angular coordinate only, hence again both expressions must equal a constant. Let this constant be m^2. Therefore

$$\frac{1}{\Phi} \frac{d^2 \Phi}{d\phi^2} = -m^2 \qquad 6.5$$

and $\quad \dfrac{1}{\Theta} \dfrac{d^2 \Theta}{d\theta^2} + \dfrac{1}{\Theta} \cot \theta \, \dfrac{d\Theta}{d\theta} + l(l + 1) - \dfrac{m^2}{\sin^2 \theta} = 0. \qquad 6.6$

We therefore conclude that, assuming the variables to be separable, the original Schrödinger equation 6.2 can be replaced by the three equations 6.3, 6.5 and 6.6, each of these involving only one coordinate. We now proceed to the consideration of the solutions of these equations.

Equation 6.5, involving the azimuthal angle, is the well-known equation for simple oscillations. It has the general solution

$$\Phi(\phi) = A \, e^{i(m\phi + B)}.$$

In order that this solution be single valued (i.e. $\Phi(\phi) = \Phi(\phi + k2\pi)$, where k is any integer) m must be integral or zero.

Consider now equation 6.6. Let $\mu = \cos \theta$; then in terms of μ this equation can be written

$$(1 - \mu^2) \frac{d^2 \Theta}{d\mu^2} - 2\mu \frac{d\Theta}{d\mu} + \left[l(l + 1) - \frac{m^2}{1 - \mu^2} \right] \Theta = 0. \qquad 6.7$$

We first consider the special case where $m = 0$ and this equation becomes

$$(1 - \mu^2) \frac{d^2\Theta}{d\mu^2} - 2\mu \frac{d\Theta}{d\mu} + l(l+1) \Theta = 0. \qquad \textbf{6.8}$$

Solutions for this equation, known as Legendre's equation, may be sought in the form of a power series in μ. In general, there will be two independent solutions. One of these is a series consisting of the odd powers of μ, the other a series

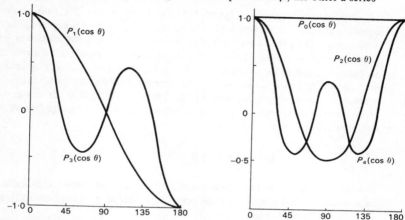

Figure 21 Form of the first few Legendre polynomials

consisting of the even powers. In the general case, no restriction being placed on the value of l, the series do not terminate after a finite number of terms and lead to infinite values of the solution when $\mu = 1$ (i.e. on the z or polar axis). They are therefore not physically acceptable solutions. However, if $l = 0$ or a positive integer, one or other of the series (depending on whether l is even or odd) terminates and leads to a solution which remains finite for the whole range of possible values of μ. These physically acceptable solutions which occur for zero or positive integral values of l are called Legendre polynomials and are denoted by $P_l(\mu)$ or $P_l(\cos\theta)$. As can be shown by finding power-series solutions of equation **6.8**, the first few Legendre polynomials are

$$P_0(\cos\theta) = 1, \qquad P_1(\cos\theta) = \cos\theta, \qquad P_2(\cos\theta) = \tfrac{1}{2}(3\cos^2\theta - 1),$$

$$P_3(\cos\theta) = \tfrac{1}{2}(5\cos^3\theta - 3\cos\theta), \qquad P_4(\cos\theta) = \tfrac{1}{8}(35\cos^4\theta - 30\cos^2\theta + 3).$$

Turning now to the case where m is not zero but a positive or negative integer, then, providing $|m| \leqslant l$, a solution of equation **6.7** which remains finite for all values of μ is

$$P_l^m(\mu) = (1 - \mu^2)^{|m|/2} \frac{d^{|m|}}{d\mu^{|m|}} P_l(\mu).$$

$P_l^m(\mu)$ is called the *associated Legendre function*.

We now consider the third equation, the radial equation 6.3. We introduce the modified radial wave function $G(r)$ defined by the equation $G(r) = r R(r)$. In terms of G, equation 6.3 can be written

$$\frac{d^2G(r)}{dr^2} + \frac{2M}{\hbar^2}\left[W - V(r) - \frac{l(l+1)\hbar^2}{2Mr^2}\right]G(r) = 0. \qquad 6.9$$

This modified radial equation is seen to be similar to the one-dimensional wave equation discussed in section 3.10. The solutions there found however are not acceptable in the present problem, as, when substituted into the relation $R = G/r$, they lead to R becoming infinite at the origin, that is, when $r = 0$. To arrive at acceptable solutions we return to equation 6.3. As the potential is being assumed constant within the well, we can, as in the one-dimensional case, introduce a constant k, the wave number, defined by $k^2 = [W - V(r)] 2M/\hbar^2$. If now $\rho = kr$ be introduced as the variable and the function R be replaced by $\sqrt{(\pi/2kr)}R'$, equation 6.3 becomes

$$\rho^2 \frac{d^2R'}{d\rho^2} + \rho \frac{dR'}{d\rho} + [\rho^2 - (l + \tfrac{1}{2})^2] R' = 0.$$

This is known as Bessel's equation and the solutions in the theory of functions are known as Bessel functions. Thus, reverting to our original function $R(r)$, a solution can be written as

$$R(r) = \sqrt{\left[\frac{\pi}{2kr}\right]} J_{l+\frac{1}{2}}(kr),$$

where $J_{l+\frac{1}{2}}(kr)$ is the Bessel function of order half an odd integer. The solution can be expressed as

$$R(r) = j_l(kr),$$

where $\quad j_l(kr) = \sqrt{\left[\frac{\pi}{2kr}\right]} J_{l+\frac{1}{2}}(kr).$

$j_l(kr)$ is called the *spherical Bessel function.*

$$j_0(kr) = \frac{1}{kr} \sin kr,$$

$$j_1(kr) = \frac{1}{(kr)^2} \sin kr - \frac{1}{kr} \cos kr.$$

Higher orders can be found by using the recurrence formula

$$j_{l+1}(kr) = \frac{2l+1}{kr} j_l(kr) - j_{l-1}(kr). \qquad 6.10$$

6.5 Orbital and magnetic quantum numbers

The formal solutions of the differential equations have been given in some detail to indicate how the restrictions on l and m values arise mathematically. The physical meaning to be assigned to these quantities has now to be discussed.

We note that the expression

$$\frac{l(l+1)\hbar^2}{2Mr^2},$$

which appears in equation **6.9**, can be written as

$$\frac{l(l+1)\hbar^2}{2\mathscr{I}},$$

where $\mathscr{I} = Mr^2$ is the moment of inertia of the single nucleon about the origin. In this form the expression is seen to be the rotational kinetic energy of the nucleon providing $\sqrt{[l(l+1)]}\hbar$ is taken as the angular momentum of the nucleon. For this reason l is termed the angular-momentum or *orbital quantum number*.

If now a particular axis is given physical significance (for example by applying a magnetic field to the system) then the angular-momentum vector $\sqrt{[l(l+1)]}\hbar$ precesses about the specified direction oriented in such a way that the component of angular momentum along the direction is $m\hbar$, where m is an integer. There are then $2l+1$ different orientations possible, corresponding to the $2l+1$ values of m ranging from $+l$ to $-l$ and including zero. m from the vector model satisfies the same conditions as the separation constant in equation **6.5**. It is identified with this separation constant and referred to as the *magnetic quantum number*.

6.6 The radial quantum number

We now consider the radial wave function $R(r)$. We take first the case where $l = 0$. It follows that a solution of **6.3** is

$$R(r) = j_0(kr) = \frac{\sin kr}{kr}.$$

This solution has the required property of remaining finite at the origin, corresponds to a standing wave inside the well and will represent a stationary state if its value is zero at the nuclear boundary. For this condition to be satisfied k must have a value given by

$$\frac{\sin kR}{kR} = 0$$

(in this equation R of course represents the nuclear radius and is not to be confused with the radial wave function). The smallest of the many values of k which satisfy this condition is given by $k_{10} R = \pi$. We note that in this case the wave function has no nodes inside the well. We recall from the definition of k that

93   The radial quantum number

$$\frac{\hbar^2 k_{10}^2}{2M} = [W_{10} - V(r)].$$

The potential in the well being constant, $\hbar^2 k_{10}^2/2M$ can be taken as a measure of the energy of the system when the nucleon is in this particular state. We now introduce a radial quantum number ν, numerically equal to one plus the number of radial nodes of the wave function within the nucleus. We use the traditional spectroscopic notation of s, p, d, f, etc. for $l = 0, 1, 2, 3$, etc. and add ν as a prefix to describe the state. In this notation, the case we have been discussing is that of a nucleon occupying a 1s state.

We note that the next highest value of k producing the correct boundary conditions, l still being zero, is given by $k_{20} R = 2\pi$. The wave function will then have one radial node corresponding to $k_{20} r = \pi$ and the radial quantum number will now therefore be equal to two. The state is a 2s state. The argument can obviously be extended to higher ν-values, l of course still being zero.

For $l = 1$ we take the next order Bessel function and find $R(r)$ to have the form

$$\frac{\sin kr}{(kr)^2} - \frac{\cos kr}{kr}.$$

The values of k which will produce nodes at the nuclear surface are now given by $\tan kR = kR$. The solutions of this equation can be obtained graphically and are $kR = 4\cdot50, 7\cdot70, 10\cdot9, \ldots$ These correspond to the 1p, 2p, 3p, $\ldots$ states.

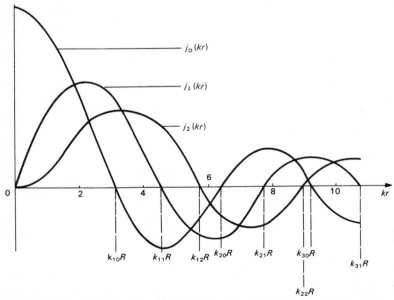

Figure 22 Plot of the first few spherical Bessel functions. The zeros give the $k_{\nu l} R$ values

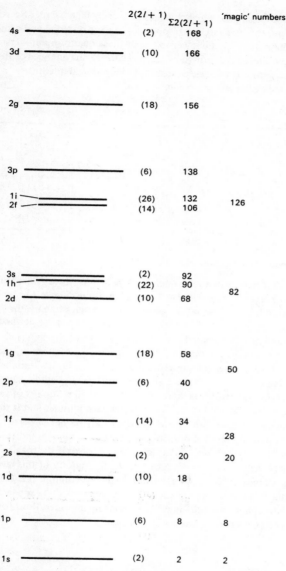

	$2(2l+1)$	$\Sigma 2(2l+1)$	'magic' numbers
4s	(2)	168	
3d	(10)	166	
2g	(18)	156	
3p	(6)	138	
1i	(26)	132	126
2f	(14)	106	
3s	(2)	92	
1h	(22)	90	82
2d	(10)	68	
1g	(18)	58	
			50
2p	(6)	40	
1f	(14)	34	
			28
2s	(2)	20	20
1d	(10)	18	
1p	(6)	8	8
1s	(2)	2	2

Figure 23 Shell-model states for rectangular potential well. The occupation numbers given in the first of the three columns on the right of the diagram are added in the neighbouring column to give the total number of nucleons accommodated up to a given level. The magic numbers above 20 do not find a place in the scheme

The argument can be extended to higher values of l, using higher-order Bessel functions, and kR values can be found corresponding to pairs of radial and orbital quantum numbers. The results for the lower values of these quantum numbers are given in Table 3 and illustrated in Figure 22. In Figure 23 the energies of the states are drawn in the conventional *level diagram*, the associated quantum numbers being indicated at the left-hand side.

Table 3

	s ($l = 0$)				p ($l = 1$)		
ν	1	2	3	4	1	2	3
$k_{\nu l} R$	3·14	6·28	9·42	12·57	4·49	7·72	10·90

	d ($l = 2$)			f ($l = 3$)		g ($l = 4$)		h ($l = 5$)	i ($l = 6$)
ν	1	2	3	1	2	1	2	1	1
$k_{\nu l} R$	5·76	9·10	12·32	6·98	10·41	8·18	11·71	9·36	10·51

6.7 The number of nucleons in the various shells

It is now postulated that in a given nucleus, having N neutrons, these N neutrons fill the lowest available levels in this scheme. The availability is determined by the Pauli exclusion principle which does not permit two particles to have the same set of four quantum numbers, ν, l, m, m_s, where ν, l and m are as defined in the previous section and m_s is associated with the component of the spin (i.e. intrinsic angular momentum) of the nucleon along a specified direction; m_s has the two values $+\frac{1}{2}$ and $-\frac{1}{2}$. Each level can therefore only contain a limited number of particles. For a given ν and a given l there are $2l + 1$ different possible values of m and two different values of m_s. Thus there are $2(2l + 1)$ different pairs of values of m and m_s available. This quantity is called the occupation number and is shown in brackets to the right of each level in Figure 23. The total number of particles accommodated in the scheme, up to and including a particular level, is indicated to the right of the occupation number of that level in each case. If the magic numbers indeed reflect shell behaviour, then the test of the validity of this level scheme will be the existence of the magic numbers in the last column. It is seen that the first three magic numbers are included but none of the others appear. There are other reasons connected with angular-momentum considerations for believing that this predicted order of levels is not correct.

6.8 Spin–orbit coupling

The suggestion which revitalized the shell model was that a coupling be assumed to exist between the orbital angular momentum and the spin angular momentum of a nucleon. Spin–orbit coupling had been found to be a feature of the atomic system and to play a fundamental and necessary role in determining the details of atomic spectra. In the atom it may be considered to arise from the interaction of the magnetic dipole moment of the charged, spinning electron with the magnetic field arising from the relative motion of the electron and the charged nucleus. In the nuclear context, however, there is no simple reason for such a coupling to be expected. Its introduction was proposed in a spirit of empiricism.

In the absence of spin–orbit coupling, l and s orient with respect to a specified direction Oz so that the observable components of angular momentum parallel to Oz are $m\hbar$ and $m_s\hbar$. When there is spin–orbit coupling, l and s form a resultant angular momentum j. There are two possible orientations of the spin angular momentum s with respect to the orbital angular momentum l, just as there are two possible orientations of s in an externally applied field. The two orientations give rise to two possible values of j, $j = l + s$ and $j = l - s$. The absolute value of the resultant angular momentum j is $\sqrt{[j(j + 1)]}\,\hbar$ and j orients with respect to Oz to give an observable value of angular momentum $m_j\hbar$ parallel to Oz, where m_j is half-integral and $|m_j| < j$. l and s can be pictured as precessing about j, which in turn, as depicted in Figure 24, precesses about Oz.

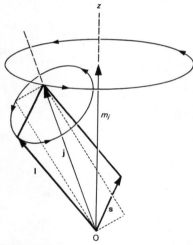

Figure 24 Vector diagram to illustrate the coupling of the orbital angular momentum l with the spin angular moment s to form j with its projection m_j

The four quantum numbers of the nucleon in the absence of spin–orbit coupling are l, s, m_l and m_s. When there is spin–orbit coupling these are replaced by l, s, j and m_j.

We note that there are $2l + 2$ values of m_j associated with the larger of the two j-values (i.e. $j = l + \frac{1}{2}$) ranging from $l + \frac{1}{2}$ to $-(l + \frac{1}{2})$ and $2l$ values of m_j associated with the smaller of the two j-values ranging from $l - \frac{1}{2}$ to $-(l - \frac{1}{2})$, giving $2(2l + 1)$ values in total. This is of course equal to the total obtained by taking the $2l + 1$ values of m_l each in association with two values of m_s. However, spin–orbit coupling implies that the $2l + 2$ states associated with $j + s$ will have a different energy from the $2l$ states associated with $j - s$, whereas in the absence of spin–orbit coupling there is complete degeneracy.

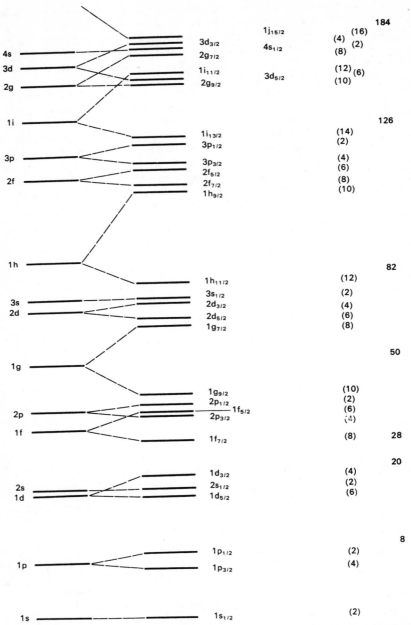

Figure 25 Modifications to the level scheme for a rectangular well brought about by the introduction of spin–orbit coupling. All the magic numbers now find a place and correspond to significantly large gaps between levels

When the hypothesis was first suggested there was no theoretical guidance as to which of the two groups of states had the greater energy or as to the energy difference arising from spin–orbit coupling. The following choice was made on the basis of its leading to a satisfactory level scheme. Firstly, it was assumed that the levels corresponding to $j = l + \frac{1}{2}$ lie lower than those corresponding to $l - \frac{1}{2}$. This is contrary to the behaviour of the electron in the atomic case, where there is a well understood magnetic spin–orbit interaction. Secondly it is assumed that the energy difference between the two sets of levels (i.e. the splitting) is proportional to l. This is in agreement with the results of the atomic system. Figure 25 shows the effect of this spin–orbit splitting on the level scheme. The occupation numbers are shown as before. It is seen now that all of the magic numbers find a place in the last column. Further they can be made to correspond to comparatively large energy gaps in the level scheme and thus to give plausibility to the idea of nucleon shells.

The level scheme in Figure 25 will be valid for nucleons of one kind in a nucleus of given A-value. For any other A-value, the nuclear radius will be different and hence the absolute position of the levels will be altered. The order of the levels however remains unchanged. We have ignored Coulomb effects and hence our result is directly applicable only to neutrons. In the case of protons, the addition of the repulsive Coulomb forces leads to levels which, on an absolute scale, are higher than the neutron levels but are in other respects similar.

6.9   Effect of shape of the nuclear well

We have assumed a highly idealized shape for the nuclear well in the discussion of section 6.4. Other shapes lend themselves to exact mathematical analysis and enable us to see the extent to which the level scheme is dependent on details of nuclear shape.

A potential of the form

$$V(r) = V_0 \left[\frac{1}{2} \left(\frac{r}{R} \right)^2 - 1 \right]$$

corresponds to a harmonic oscillator. Its energy levels are evenly spaced and there are degeneracies which disappear when we go to the other extreme of the rectangular well.

A more realistic intermediate shape has received considerable attention (Woods and Saxon, 1954); the potential shape is given by

$$V(r) = \frac{-V_0}{1 + e^{(r-R)/a}}.$$

The shape of this well is compared with the two extremes in Figure 20 and the corresponding energy levels are shown in Figure 26.

We conclude that, although the energy intervals are affected, the order of states (apart from the 1h and 1i states) is not sensitive to the degree of flatness of the well or steepness of the potential rise.

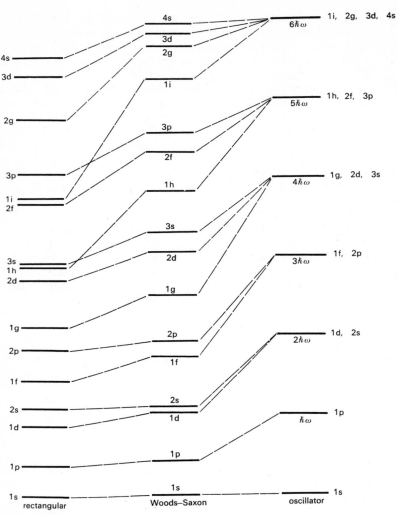

Figure 26 Effect of 'flatness' of well and steepness of sides of well on the level diagram. With the exception of the 1h and 1i states the order of states is not affected

6.10 Nuclear ground-state spins

The shell-model level scheme will now be used to interpret the observed spins (i.e. intrinsic angular momenta) of nuclear ground states.

The nucleons of both kinds will in the ground state (i.e. the state of lowest energy) be in the lowest levels available to them. It is to be expected that when a level is fully occupied the summed contributions of the individual nucleons in the level to the total angular momentum will be zero. When a level is partly filled by an even number of nucleons, it is found experimentally that the nucleons pair off in such a way that the total angular momentum is still zero. We can state this with certainty as there is no exception known to the rule that the measured ground-state spin of an (even, even) nucleus is zero. When the partly filled level contains an odd number of nucleons we assume that again all nucleons, except the last, pair off in a way which leads to cancellation of their contribution to the total spin, and that the spin of the whole nucleus in the ground state is then given by the angular momentum of the single unpaired nucleon. On this assumption we can then predict, from the scheme in Figure 25, the ground-state spin of (odd, even) or (even, odd) nuclei.

For example, $^{25}_{12}$Mg will have a spin determined by the thirteenth neutron. This nucleon is seen to be in a $1d_{\frac{5}{2}}$ state and hence the predicted spin value is $\frac{5}{2}$. Again, $^{69}_{31}$Ga will have a spin determined by the thirty-first proton. This is seen to be in a $2p_{\frac{3}{2}}$ state and we would thus predict a spin of $\frac{3}{2}$. Predictions made in this way with only a few exceptions (see nuclides marked † in Appendix A) are in agreement with the observed values. The general pattern of spin values provides the angular-momentum evidence for magic numbers mentioned above. For example, as we go from light to heavy nuclei the first time that a spin as high as $\frac{9}{2}$ is encountered is when we reach $^{73}_{32}$Ge, and we note that the forty-first neutron is the first neutron in the $1g_{\frac{9}{2}}$ level. If we confine attention to (odd, even) nuclei, the first time a spin of $\frac{9}{2}$ is encountered is when $^{93}_{41}$Nb is reached, the forty-first proton being the first proton in the $1g_{\frac{9}{2}}$ level.

The direct predictions break down when high spin values are involved. For example, the seventy-first neutron should be the first to occupy the $1h_{\frac{11}{2}}$ level and from there until the eighty-second neutron is reached (even, odd) nuclei would be expected to have spins $\frac{11}{2}$. $^{123}_{52}$Te has however a measured spin of $\frac{1}{2}$ as has $^{129}_{54}$Xe, while $^{131}_{54}$Xe, $^{135}_{56}$Ba and $^{137}_{56}$Ba all have spins $\frac{3}{2}$. Attempts to explain this have been made with some success by considering the pairing energy discussed in section 5.5 to be greater the higher the l-values of the two nucleons concerned. If this is correct, then an unpaired nucleon in an h-state would be expected to split a pair of nucleons in an s-state, pairing with one and leaving the other unpaired, providing the difference in the pairing energy for the h- and s-state nucleons exceeds the energy necessary to raise a nucleon from the s to the h energy level.

In the case of (odd, odd) nuclei there will be two unpaired nucleons, one of each kind, to consider. There is nothing in the model to predict how their angular momenta will couple. We saw in section 5.4 that there are only four examples of stable nuclei in this category; in each case the nuclear spin is less than the sum of the j-values of the two unpaired nucleons. This of course must be so whatever the details of the coupling if the behaviour is to be compatible with the shell model. There are one or two examples of unstable nuclei in the (odd, odd) category having long enough half-life to permit determinations of their spins to be made by experimental methods. The results in these cases too are compatible with compounding of the *j*-values of the two unpaired nucleons. An interesting case is $^{50}_{23}$V, whose gound state has the high spin value of 4. The twenty-third proton and twenty-seventh neutron are both in $1f_{\frac{7}{2}}$ states and can therefore combine to produce the high value of spin measured.

There are one or two comparatively rare instances of the prediction breaking down among the light nuclei. For example, $^{21}_{10}$Ne would be expected to have a spin determined by the eleventh neutron, which is in the $1d_{\frac{5}{2}}$ level. Its measured spin is however $\frac{3}{2}$. It has to be conjectured, therefore, that in this case the pairing is broken and the three neutrons in the $d_{\frac{5}{2}}$ level compound their spins to produce a spin of $\frac{3}{2}$.

6.11 Islands of isomerism

Usually, when a nuclear excited state de-excites by the emission of electromagnetic radiation (i.e. γ-ray emission), the transition probability for the process leads to half-lives of the order of 10^{-16} s. We shall see in Chapter 10 that if a very large difference in angular momentum exists between the initial and final states the process can be very much inhibited. In these circumstances the half-life can be very long indeed. For example, there is a state $^{110}_{47}$Ag which has a half-life for de-excitation of 253 days. When the excited state is long enough lived, the specimen will constitute a γ-source decaying exponentially with time. Apart from the γ-ray there is no other product of de-excitation of the state (see however section 10.2 and Figure 63). The state is said to be an *isomer* and the de-excitation is referred to as an *isomeric transition*. The range of half-lives accessible to experimental measurement has been extended to lower and lower values as electronic techniques have developed. Lives shorter than picoseconds (10^{-12} s) have now been convincingly measured, as we shall discuss in Chapter 10. Although strictly speaking these are isomers as defined above, the term is usually kept for states with a half-life of a microsecond or longer.

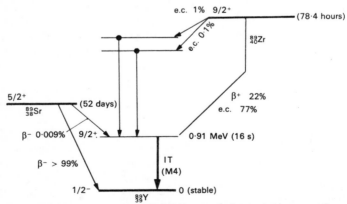

Figure 27 Decay scheme to show the populating and the properties of a $\frac{9}{2}^{+}$ isomeric state of ^{89}Y

Within the framework of the shell model, some at least of the excited states of a nucleus can be visualized as arising from a promotion of one nucleon from the topmost level filled, or partly filled, to an unoccupied level of higher energy. Should there be, closely above the topmost occupied level, a level of much higher angular momentum, then the conditions necessary for isomeric behaviour may exist. A good example of this is the stable nucleus $^{89}_{39}$Y. The ground-state spin of this nucleus is dictated by the thirty-ninth proton, which is in the $2p_{\frac{1}{2}}$ state. Close above this is the unoccupied $1g_{\frac{9}{2}}$ state. The promotion of the unpaired proton from the p-state to the g-state gives rise to an excited state of the nucleus, shown in Figure 27, 0·91 MeV above the ground state. The ^{89}Y nucleus can be left in this state following the β^- decay of ^{89}Sr or the β^+ (or electron-capture) decay of ^{89}Zr. When so formed the ^{89}Y nucleus in decaying to its ground state is involved in a change of $4\,\hbar$ in angular momentum. As a consequence, the half-life for this electromagnetic transition is observed to be sixteen seconds. A further example of isomeric decay is to be found in the level scheme of ^{87}Y illustrated in Figure 28.

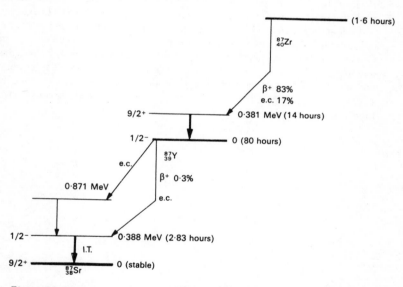

Figure 28 Decay scheme to show the populating and properties of an isomeric state in ^{87}Y and an isomeric state in ^{87}Sr

There is an excited state of the same nature as that found in ^{89}Y. In this case the excited state lies closer in energy to the ground state, and the half-life (14 hours) as a consequence is even longer. Further, note that ^{87}Y decays to yet another isomer, in this case a state of ^{87}Sr. The ground state of this nucleus has a spin of $\frac{9}{2}$ due to the unpaired forty-ninth neutron. There is a low-lying excited state of spin $\frac{1}{2}$, which can be interpreted as arising from a $2p_{\frac{1}{2}}$ neutron in the state below being promoted to a vacancy in the $g_{\frac{9}{2}}$ state. Once again the large difference in angular momentum explains the isomeric decay.

Thus in a very satisfactory way the position of the islands of isomerism on the nuclear chart find an explanation within the framework of the shell model. It should be noted that all excited states do not have such simple configurations as we have been picturing above. Those states which are as simple as this are referred to as *single-particle states*. However, it is clear that as well as single-particle states it must also be expected that there will exist excited states involving the promotion of two or even more particles.

6.12   Summary

The shell model, originating in an attempt to meet the challenge of explaining magic numbers and based on the arbitrary assumption of spin–orbit coupling, produces a level scheme which enables quantum numbers to be assigned to nucleons in complex nuclei. With the additional assumption that nucleons pair so as to cancel angular momentum, a scheme of ground-state spins in excellent agreement with measured ground-state spin values for odd-A nuclei can be constructed. In some cases the spins of excited states of nuclei can be explained on the basis of one particle being promoted from the lowest energy configuration and in particular the existence of islands of isomerism has a simple interpretation within the shell model. We shall see in the following chapters that the shell model has also a role to play in the interpretation of the measured electric and magnetic moments of nuclei.

Chapter 7
Nuclear Moments 1

7.1   Single-particle magnetic dipole moment

We saw in section 6.9 that cancellation of angular momentum takes place as long as there is an even number of nucleons in the nucleus; when there is an odd number of nucleons the total spin of the nucleus can be identified with the angular momentum of the one unpaired nucleon. Now it is found that (even, even) nuclei, which, as seen above, invariably have zero spin, also have zero magnetic dipole moment. It would therefore seem instructive to consider whether the observed finite magnetic dipole moments of odd-A stable nuclei can be explained as arising from the motion of the single unpaired nucleon. We now proceed to consider the magnetic moment to be expected to arise from the single nucleon in a given quantum state.

First however we have recourse to classical physics to define the basic terms used in atomic and nuclear magnetic studies. A particle with electric charge e, measured in electrostatic units, and mass M, moving in a circular orbit of radius r with constant angular velocity ω, is equivalent to a current $e\omega/2\pi c$ flowing so as to enclose an area πr^2. The equivalent magnetic dipole moment from classical electromagnetism is $e\omega r^2/2c$. We note that the angular momentum of the particle about an axis normal to the orbit and through its centre is $Mr^2\omega$. Thus the ratio of magnetic moment to angular momentum, the *gyromagnetic ratio*, is $e/2Mc$. Denoting the magnetic dipole moment by μ and the angular momentum by I we thus have

$$\frac{\mu}{I} = \text{constant} = \gamma, \qquad\qquad 7.1$$

say, where γ is independent of ω and r. This simple result for a circular orbit is true in the general case, where the orbit may be elliptic. In the quantum-mechanical case, the maximum value of angular momentum along a specified direction for a particle in an l-state is $l\hbar$ and the measured associated dipole moment would therefore be expected to be $le\hbar/2Mc$. If we define the constant quantity $e\hbar/2Mc$ to be the standard unit of magnetic moment, then the magnetic moment in terms of this standard unit is numerically equal to l. In the case of the electron, $e\hbar/2Mc$ is termed the *Bohr magneton*. When we apply the same ideas to the nucleon, the analogous quantity is termed the *nuclear magneton*. The larger value of the nucleon mass means that the nuclear magneton is $1/1836$ times the Bohr magneton. Thus on the whole, since the angular momenta in both cases is of the order of $\hbar$, nuclear magnetic moments are expected to be smaller by this factor than atomic magnetic moments. When the accepted values of the constants are inserted in the expression $e\hbar/2Mc$, the nuclear magneton is found to be $5 \cdot 050 \times 10^{-31}$ J G^{-1}.

We now consider the magnetic moment which classical physics attributes to a rotating charged body. Assume that the charge is distributed through the volume to give a charge density ρ_e, which may vary from point to point. Let the mass be distributed in such a way as to give a mass density ρ, which also may vary from point to point. Take an element of the body of volume dV at a distance $\bar{r}$ from the axis of rotation. This element is equivalent to an orbiting particle having a charge $\rho_e\,dV$ and a mass $\rho\,dV$. It will therefore contribute to the magnetic moment an amount

$$d\mu = (\rho\,dV)\,\bar{r}^2\omega\left[\frac{\rho_e\,dV}{2\rho\,dV\,c}\right],$$

where ω is the angular velocity. If ρ_e and ρ are constants, or have the same dependence on the space coordinates, then we can write

$$\rho_e = K\rho,$$

where K is a constant. The total charge

$$e = \int \rho_e\,dV = K\int \rho\,dV = KM,$$

where M is the total mass. In this special case therefore

$$\frac{\rho_e}{\rho} = K = \frac{e}{M}.$$

Integrating over the whole body we then have

$$\mu = \frac{e\omega}{2Mc}\int \rho\bar{r}^2\,dV = \frac{e}{2Mc} \times \text{angular momentum}.$$

If on the other hand ρ_e and ρ do not have the same dependence on the space coordinates (i.e. the mass and charge distributions are dissimilar) then

$$\mu = \frac{\omega}{2c}\int \bar{r}^2\rho_e\,dV = \text{angular momentum} \times \frac{1}{2c}\,\frac{\int \bar{r}^2\rho_e\,dV}{\text{moment of inertia}}.$$

As an example, consider a sphere, radius R, of constant density, with the charge located entirely in a surface layer of thickness t, where $t \ll R$. Taking a ring element,

$$dV = 2\pi\bar{r}tR\,d\theta,$$

where θ is the usual spherical polar coordinate, and

$$\int \bar{r}^2\rho_e\,dV = \int\limits_0^\pi 2\pi R^4 t\rho_e \sin^3\theta\,d\theta = \tfrac{2}{3}eR^2,$$

since e, the total charge, equals $4\pi R^2 t\rho_e$.

Thus $\quad \mu =$ angular momentum $\times \dfrac{1}{2c} \times \dfrac{\frac{2}{5}eR^2}{\frac{2}{5}MR^2} =$ angular momentum $\times \dfrac{5}{6}\dfrac{e}{Mc}$.

Therefore $\quad \dfrac{\mu}{I} = \dfrac{e}{2Mc}\dfrac{5}{3}$.

We thus see that in the general case a numerical coefficient is involved which is dependent on the charge and mass distributions. We therefore modify equation 7.1 by introducing the so called g-factor, writing

$$\frac{\mu}{I} = g\gamma. \tag{7.2}$$

In the case of the electron, Dirac's theory leads to a predicted value for g of 2, and this is in good agreement with the electron's measured magnetic dipole moment. In the case of the nucleon, there is no similar theoretical guidance and we use the measured values of μ together with equation 7.2 to arrive at the value of g. For the proton, $\mu_p = 2{\cdot}7934$ nuclear magnetons, and therefore the g-factor, which for rotational or spin angular momentum we shall denote by g_{sp}, will be $\mu_p/\frac{1}{2} = 5{\cdot}5868$. It is found experimentally that the neutron, despite its having zero net electric charge, has a finite magnetic dipole moment. This indicates that the neutron has internal electrical structure with different positive and negative charge distributions, the total charge being zero. The measured magnetic moment in the case of the neutron is $-1{\cdot}9135$ nuclear magnetons, the negative sign indicating that the direction of the dipole is related to the spin direction as it would be for a negatively charged body. The corresponding g-factor, $g_{sn} = \mu_n/\frac{1}{2} = -3{\cdot}8270$. It is convenient to modify equation 7.1 when applied to orbital motion by introducing in this case too a g-factor. For the proton, which behaves in this respect as a classical point charge, the value of this g-factor g_{lp} will be unity. The neutron, again as would be expected on the classical view, makes no contribution to the magnetic moment by virtue of its orbital motion and therefore $g_{ln} = 0$.

7.2 Relationship of magnetic moment to nuclear spin

Let the nucleon be in an l-state. Then the angular-momentum vector diagram when spin–orbit coupling is assumed is drawn in Figure 29. The orbital angular momentum of absolute magnitude $\sqrt{[l(l+1)]}\,\hbar$ we denote by $\mathbf{l}$, the spin angular momentum by $\mathbf{s}$ and their resultant by $\mathbf{j}$, where, as in section 6.7, $\mathbf{j} = \mathbf{l} + \mathbf{s}$. The vectors $\mathbf{l}$ and $\mathbf{s}$ precess about $\mathbf{j}$, which in turn precesses about the direction of an applied magnetic field. $\mathbf{j}$ has a component $j\hbar$ along the field direction. If the g_l and g_s factors were equal then the same diagram, suitably scaled, would represent the magnetic dipole-moment vectors. However we have seen that the g_l and g_s factors are not equal for either type of nucleon. The magnetic moment associated with $\mathbf{l}$ will have a component along the direction of $\mathbf{j}$ and also a component perpendicular to it. The perpendicular component will time-average

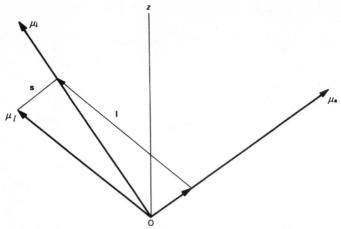

Figure 29 Contributions to nuclear magnetic dipole moment, arising from nucleon orbital and spin angular momenta, plotted on an angular-momentum vector diagram. Note that μ_j is only one component of the resultant of μ_l and μ_s

to zero. Similarly for the components associated with s. If we then take the sum of the components along j and resolve this sum along and perpendicular to the field direction, again the perpendicular component can be discounted and the measured dipole moment is the component of the sum along the field direction, when the angle between j and the field is the smallest of the discrete number of permitted angles. Carrying out this programme we have, for the sum of the components of the moments along j, the expression

$$\frac{1}{\hbar} \left[g_l |\mathbf{l}| \cos(\mathbf{l}, \mathbf{j}) + g_s |\mathbf{s}| \cos(\mathbf{s}, \mathbf{j}) \right].$$

Resolving this along the field direction, the observed moment is found to be given by μ, where

$$\mu = \frac{1}{\hbar} \left[g_l |\mathbf{l}| \cos(\mathbf{l}, \mathbf{j}) + g_s |\mathbf{s}| \cos(\mathbf{s}, \mathbf{j}) \right] \cos(\mathbf{j}, Oz)$$

$$= \frac{g_l \left[l(l+1) + j(j+1) - s(s+1) \right] + g_s \left[s(s+1) + j(j+1) - l(l+1) \right]}{2(j+1)},$$

using the elementary trigonometrical formula for the cosine of the angle in a triangle.

Take now the case $j = l + s$. We substitute $\frac{1}{2}$ for s and $j - \frac{1}{2}$ for l, and find

$$\mu_{l+\frac{1}{2}} = g_l(j - \frac{1}{2}) + \frac{1}{2}g_s = lg_l + \frac{1}{2}g_s.$$

For $j = l - s$, we have $s = \frac{1}{2}$ and $l = j + \frac{1}{2}$, and therefore

$$\mu_{l-\frac{1}{2}} = \frac{j}{j+1} \left[g_l(j + \frac{3}{2}) - \frac{1}{2}g_s \right] = \frac{j}{j+1} \left[(l+1)g_l - \frac{1}{2}g_s \right].$$

We must treat the cases of the proton and neutron separately, as their g-factors are different. If the particle concerned is a proton ($g_{lp} = 1$, $g_{sp} = 5\cdot5863$) and we take the case $j = l + \frac{1}{2}$, then $\mu_p = j + 2\cdot29$.

For a proton and $j = l - \frac{1}{2}$,

$$\mu_p = \frac{j}{j+1}(j - 1\cdot29)$$

$$= j - 2\cdot29\frac{j}{j+1}.$$

For a neutron ($g_{ln} = 0$, $g_{sn} = -3\cdot8270$) and $j = l + \frac{1}{2}$,

$$\mu_n = -1\cdot91.$$

For a neutron and $j = l - \frac{1}{2}$,

$$\mu_n = 1\cdot91\frac{j}{j+1}.$$

7.3 Schmidt lines

If we now use these results in conjunction with the shell-model hypothesis (namely the assumption that the spin and magnetic moment in odd-A nuclei

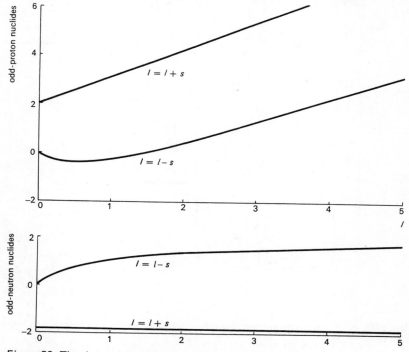

Figure 30 The theoretical Schmidt lines of magnetic dipole moment plotted against nuclear spin quantum number

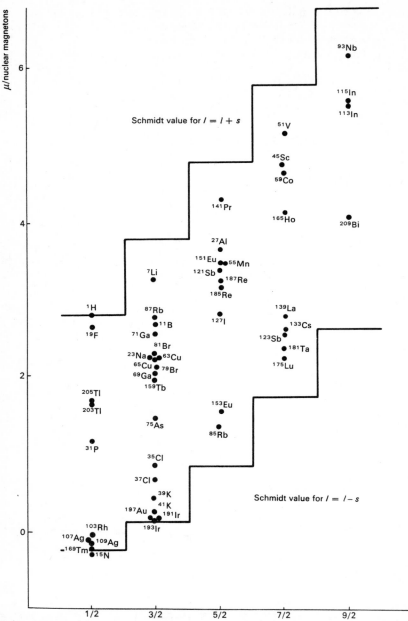

Figure 31 Plot of experimental values of nuclear magnetic dipole moment against nuclear spin quantum number of odd-*A* nuclei containing an odd number of protons

arise solely from the motion of the unpaired nucleon), then I, the nuclear spin, can be equated with j. The predicted magnetic dipole moments for odd-A nuclei will then be as in Table 4. I has, of course, only discrete values. However, if we

Table 4

	Proton (odd Z : even N)	Neutron (even Z : odd N)
$j = l + s$	$I + 2 \cdot 29$	$- 1 \cdot 91$
$j = l - s$	$I - 2 \cdot 29 \dfrac{I}{I + 1}$	$1 \cdot 91 \dfrac{I}{I + 1}$

treat it as a continuous variable for diagrammatic purposes, then the predictions for the μ-values lie on the lines shown in Figure 30. These are known as the *Schmidt lines*. In Figures 31 and 32 histograms are drawn to correspond to the discrete values of I. On the diagrams are plotted the measured dipole moments for a series of nuclei. It can be seen that the agreement is by no means perfect. However, the lines clearly set limits to the measured values. In most cases the measured value falls much closer to one line than to the other and in these cases an assignment of the nucleus to one group or the other can be made with some confidence. This is important as it provides a means of discovering, once the j-value has been determined from a measurement of I, which of the two possible l-values has to be assigned to the unpaired nucleon. This in turn permits us to establish the parity of the nucleus, positive parity arising from zero and even values of l, negative parity from odd l-values. We return in a later chapter to the discrepancy between the measured and the 'single-particle' value of magnetic dipole moment.

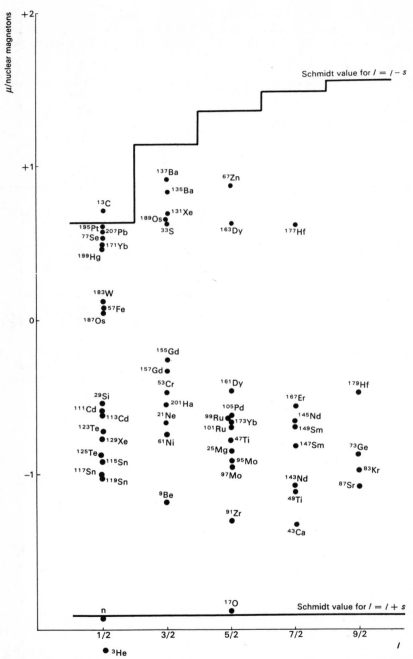

Figure 32 Plot of experimental values of nuclear magnetic dipole moment against nuclear spin quantum number for odd-A nuclei with an odd number of neutrons

7.4 Nuclear electric moments

The shell model claims to provide wave functions for the individual nucleons and hence should enable a prediction to be made of the spatial distribution of nucleons within the nuclear volume. As far as the protons are concerned, the prediction can be tested experimentally by investigating the static electric moments associated with the distribution. The measurement of these electric moments is related to the energy of the charge distribution as a function of its orientation when it is placed in an electric field. We therefore begin by discussing on the basis of classic electrostatics the energy relation between an external electric field and the internal distribution of nuclear charge.

We assume that the nucleus has a spin I and that by virtue of the rotation the averaged nuclear charge, which is the quantity measured experimentally, will be symmetric about the axis of I. We therefore limit our treatment to distributions having axial symmetry.

7.5 Multipole expansion of the electric potential

Consider $V(R, \theta)$, the electric potential at a point P with coordinates (R, θ), outside the volume of the charge distribution. As V must satisfy Laplace's equation, we can write

$$\frac{\partial^2 V}{\partial x^2} + \frac{\partial^2 V}{\partial y^2} + \frac{\partial^2 V}{\partial z^2} = 0. \qquad 7.3$$

This can be transformed into the usual spherical coordinates. As we are assuming symmetry of the charge about the z-axis, the differential coefficient with respect to ϕ will be zero. Thus the transformed equation (see equation 6.2) is

$$\left[\frac{\partial^2}{\partial R^2} + \frac{2}{R} \frac{\partial}{\partial R} + \frac{1}{R^2} \frac{\partial^2}{\partial \theta^2} + \frac{\cot \theta}{R^2} \frac{\partial}{\partial \theta} \right] V = 0. \qquad 7.4$$

If a solution with the variables separated, of the form $V(R, \theta) = F(R)G(\theta)$, be assumed, then

$$\frac{R^2}{F(R)} \frac{d^2 F(R)}{dR^2} + \frac{2R}{F(R)} \frac{dF(R)}{dR} = -\frac{1}{G(\theta)} \frac{d^2 G(\theta)}{d\theta^2} - \frac{\cot \theta}{G(\theta)} \frac{dG(\theta)}{d\theta} \qquad 7.5$$

and each side of this equation must then be independent of R and θ. Let this constant be $l(l+1)$. The right-hand side of the equation then, with $\mu = \cos \theta$, reduces to

$$(1 - \mu^2) \frac{d^2 G}{d\mu^2} - 2\mu \frac{dG}{d\mu} + l(l+1) G = 0.$$

As noted in section 6.3, if we limit the choice of l to zero or positive integral values, this equation has as a solution $G = P_l(\cos \theta)$, where $P_l(\cos \theta)$ is a Legendre polynomial.

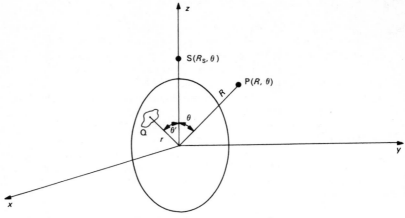

Figure 33 Charge distribution symmetrical about z-axis

Taking now the left-hand side of the equation we have

$$R^2 \frac{d^2 F(R)}{dR^2} + 2R \frac{dF(R)}{dR} = l(l+1) F(R). \qquad 7.6$$

Assuming a solution of the form $F(R) = a_n/R^n$, where n is a positive integer (thus ensuring that $F(R)$ has the physically necessary property of tending to zero as R tends to infinity), and substituting in equation 7.6, we find that

$$n(n+1) \frac{a_n}{R^n} - 2n \frac{a_n}{R^n} = l(l+1) \frac{a_n}{R^n}.$$

If $F(R)$ is to be a solution, then we must have $n = l + 1$. For the solution of equation 7.4 we therefore finally have

$$V(R, \theta) = \frac{1}{R} \sum_{l=0}^{\infty} \frac{a_l}{R^l} P_l (\cos \theta). \qquad 7.7$$

The values of the constants a_l will depend on the distribution of charge within the central volume. We can find the form of this dependence by considering the potential at the point S whose coordinates are $(R_S, 0)$. Since $P_l(1) = 1$ for all values of l, the potential at S, from equation 7.7, is given by

$$V(R_S, 0) = \frac{1}{R_S} \sum_{l=0}^{\infty} \frac{a_l}{R_S^l}. \qquad 7.8$$

The potential at S can also be expressed directly in terms of the charge distribution. If $\rho_e(r, \theta')$ is the charge density at Q, the point (r, θ') within the nucleus, and $d\tau$ is an element of volume containing the point Q, then

$$V(R_S, 0) = \int \frac{\rho_e(r, \theta')}{SQ} d\tau$$

$$= \int \frac{\rho_e(r, \theta') d\tau}{\sqrt{(R_S^2 + r^2 - 2rR_S \cos \theta')}}$$

$$= \frac{1}{R_S} \int \frac{\rho_e(r, \theta') d\tau}{\sqrt{\{1 - 2(r/R_S) \cos \theta' + r^2/R_S^2\}}} \qquad \textbf{7.9}$$

Now, as can be shown by expanding by the binomial theorem,

$$\frac{1}{\sqrt{(1 - 2x \cos \theta + x^2)}} = \sum_{n=0}^{\infty} x^n P_n (\cos \theta).$$

Hence we can express equation 7.9 as

$$V(R_S, 0) = \frac{1}{R_S} \sum_{n=0}^{\infty} \int \frac{\rho_e(r, \theta') r^n}{R_S^n} P_n(\cos \theta') d\tau. \qquad \textbf{7.10}$$

If now we compare the coefficients of $1/R_S^{l+1}$ in equations 7.8 and 7.10, we see that

$$a_l = \int \rho_e(r, \theta') r^l P_l(\cos \theta') d\tau$$

is the relationship between the coefficients and the charge distribution.

7.6   **Definition of the static electric moments**

We proceed to consider the physical meaning to be attributed to the coefficients a_l in the above multipole expansion.

Since $P_0(\cos \theta) = 1$, for all values of θ, it follows that

$$a_0 = \int \rho_e(r, \theta') d\tau = Ze.$$

The first coefficient is thus seen to equal the total charge within the nuclear volume.

Thus, if in equation 7.7 R is assumed to be so large that we need only consider the first term in the series, the potential for the given distribution is the same as that for a point charge.

Proceeding to the next term we find that, since $P_1(\cos \theta) = \cos \theta$, we have

$$a_1 = \int \rho_e(r, \theta') r \cos \theta' d\tau = \int \rho_e(r, \theta') z d\tau. \qquad \textbf{7.11}$$

This expression is the component of the first moment of the charge taken parallel to the z-axis. Note that the components of the first moment parallel to the other axes are zero because of the symmetry. The simplest distribution giving rise to a finite moment of this order and having net zero charge is the dipole formed by the displacement of equal and opposite charges e equal and opposite distances d from the origin. For this distribution $a_1 = 2ed$. At great distances from the dipole the potential as a function of θ is, to the order R^{-2}, given by

$$V = \frac{2ae}{R^2} \cos \theta,$$

in line with the angular-dependent factor in the second term of equation 7.7.

Thus, if we wish to include the second term in equation 7.7, we can regard the general charge distribution as equivalent to a point charge at the centre plus a dipole also at the centre, the dipole moment being calculated from equation 7.11. When the charges in the distribution are all of the same sign, a dipole moment can arise if the charges are not symmetrically distributed about the point with respect to which the potential is calculated. This point will normally be the centre of mass of the system. We shall see later that, in fact, nuclei do not have finite electric dipole moments.

The third term in equation 7.7 involves $P_2(\cos \theta)$, which is equal to $\frac{1}{2}(3 \cos^2\theta - 1)$. Hence we have

$$a_2 = \frac{1}{2} \int \rho_e(r, \theta') r^2(3 \cos^2\theta - 1) \, d\tau = \frac{1}{2} \int \rho_e(r, \theta')(3z^2 - r^2) \, d\tau. \qquad 7.12$$

We now introduce

$$Q = \frac{1}{e} \int \rho_e(r, \theta')(3z^2 - r^2) \, d\tau$$

and can therefore write $a_2 = \frac{1}{2}eQ$. We call Q the *quadrupole moment* and on this definition, which is that now commonly used in nuclear physics, it has the dimensions of area.

The simple distribution shown in Figure 34, which is the symmetrical displacement of equal and opposite dipoles from the origin, is seen to have zero net charge and zero dipole moment. The quadrupole moment as defined above is given by $Q = 4b^2$, since $z = r$ for this linear quadrupole. The potential at P is thus

$$V_P = \frac{eb^2}{R^3} (3 \cos^2\theta - 1),$$

to lowest order in R^{-1}.

This process of developing multipoles which are equivalent to the terms in equation 7.7 can be continued in an obvious way to higher orders. However, while electric quadrupole moments are of important current interest in nuclear physics, information about possible moments of higher order is not yet accessible to the experimentalist.

The odd electric moments – dipole, octupole, etc. – are zero for a nuclear system which has a definite parity, whether that parity be positive or negative. This is so because the amplitude of the wave function being unaltered when

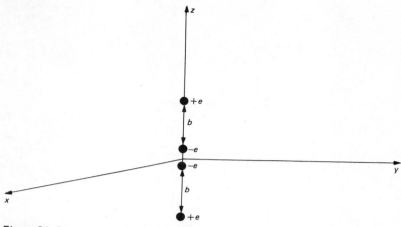

Figure 34 System of four point charges having zero net charge, zero net dipole moment and finite quadrupole moment with symmetry about the z-axis. Note that in a constant electric field the dipoles experience equal and opposite couples; the system experiences no resultant couple. However, in a field with a constant gradient parallel to the z-axis the couples on the dipoles are no longer equal, so that the system experiences a resultant couple. The net force on each dipole is equal and opposite if the field gradient is constant so that there is no resultant *force* on the system

$(-x, -y, -z)$ is substituted for (x, y, z) means that $\rho_e(r, \theta')$, which is proportional to ψ^2, is unaltered on this transformation. For odd values of l however $P_l(\cos \theta)$, the other factor in the integral defining the moment of the multipole, is an odd function in $\cos \theta$. The above parity transformation, which changes the sign of $\cos \theta$, thus changes the sign of $P_l(\cos \theta)$. Consequently when the integration is performed over the whole volume for odd values of l the result is zero. Similar arguments applied to the magnetic moments show that if the nuclear system has a definite parity, then the *even* magnetic moments – i.e. quadrupole etc. – are zero. The fact that these predictions seem to be borne out when nuclear electric and magnetic moments are measured indicates that nuclei do have a definite parity. We note that this means that their shapes must be symmetrical about the xy plane. Thus spheres and ellipsoids of revolution are permitted but a pear shape would violate the parity requirement.

7.7 Quadrupole moment of deformed sphere

In the interpretation of quadrupole moments we shall have occasion to picture nuclei as slightly deformed spheres. We now proceed to calculate the quadrupole moment of a uniformly charged sphere, slightly deformed to become an ellipsoid of revolution, having a semi-axis length c along the z-axis and semi-axes each length b along the x- and y-axes.

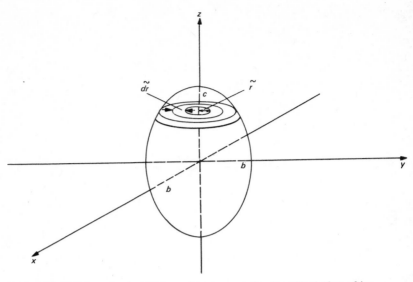

Figure 35 Division of spheroid into ring elements for the calculation of its quadrupole moment

The quadrupole moment Q will then be given by

$$eQ = \int \rho_e (3z^2 - r^2) \, d\tau$$

$$= \int \rho_e (3z^2 - z^2 - \tilde{r}'^2) 2\pi \tilde{r}' \, d\tilde{r}' \, dz,$$

where the volume element $d\tau$ has been taken as a ring lying in the disc at a distance z from the origin, as depicted in Figure 35. Hence

$$eQ = 2\pi\rho_e \int_{-c}^{+c} dz \int_{0}^{y^2} (2z^2 - \tilde{r}'^2) \tilde{r}' \, d\tilde{r}',$$

where y^2 is defined by the equation of the generating ellipse, namely

$$\frac{z^2}{c^2} + \frac{y^2}{b^2} = 1.$$

On integrating we have

$$eQ = \tfrac{8}{15} \pi \rho_e \, cb^2 (c^2 - b^2)$$

$$= Ze \tfrac{2}{5} (c^2 - b^2),$$

since the volume of the ellipsoid is $\tfrac{4}{3}\pi \, cb^2$ and Ze is the total charge. We thus have

$$Q = Z \tfrac{4}{5} \eta R^2,$$

where $\eta = \dfrac{c^2 - b^2}{c^2 + b^2}$ and $R^2 = \dfrac{c^2 + b^2}{2}.$

We note that the algebraic sign of Q depends on the relative lengths of c and b, and is positive for prolate spheroids and negative for oblate spheroids.

7.8 The interaction energy of a charge distribution with an electric field

We consider now the interaction energy of a charge distribution with an electric field. The field is assumed to be axially symmetric about an axis which we shall denote by Oz'. In the practical cases to which we shall wish to apply this theory, the field will be that at the nucleus arising from the Coulomb charge of the atomic electrons. In that circumstance Oz' will be in the direction of $\mathbf{J}$, where

$\mathbf{J} = \mathbf{L} + \mathbf{S}$,

$\mathbf{L}$ being the resultant of the orbital angular momenta of the electrons and $\mathbf{S}$ the resultant of their individual spins. We take an axis system Ox', Oy', Oz', symmetrical with respect to the field, and a system Ox, Oy, Oz symmetrical with respect to the charge distribution, as in Figure 36. The electric field we take

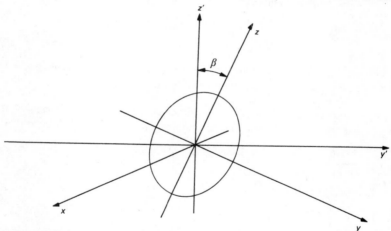

Figure 36 Ox', Oy', Oz' are fixed axes in an electric field which is symmetric about the z'-axis. Ox, Oy, Oz are axes fixed with respect to the spheroid. The x- and x'-axes instantaneously coincide

to be defined by an electrostatic potential $\phi(x, y', z')$. This function may be expanded in the neighbourhood of the origin, which we take to be the centre of mass of the nucleus and of the atom as a whole, to give

$$\phi(x, y', z') = \phi(0) + \left(x \frac{\partial\phi(0)}{\partial x} + y' \frac{\partial\phi(0)}{\partial y'} + z' \frac{\partial\phi(0)}{\partial z'} \right) +$$

$$+ \frac{1}{2} \left(x^2 \frac{\partial^2\phi(0)}{\partial x^2} + 2xy' \frac{\partial^2\phi(0)}{\partial x\,\partial y'} + \ldots + z'^2 \frac{\partial^2\phi(0)}{\partial z'^2} \right) + \ldots .$$

Taking $\rho_e(x, y', z')$ as the charge density, the energy of the system can then be expressed as

$$W = \int \rho_e(x, y', z') \, \phi(x, y', z') \, d\tau$$

$$= \int \rho_e(x, y', z') \, \phi(0) \, d\tau + \int \rho_e(x, y', z') \left[x \frac{\partial \phi(0)}{\partial x} + y' \frac{\partial \phi(0)}{\partial y'} + z' \frac{\partial \phi(0)}{\partial z'} \right] d\tau +$$

$$+ \frac{1}{2} \int \rho_e(x, y', z') \left[x^2 \frac{\partial^2 \phi(0)}{\partial x^2} + 2xy' \frac{\partial^2 \phi(0)}{\partial x \, \partial y'} + \ldots + z'^2 \frac{\partial^2 \phi(0)}{\partial z'^2} \right] d\tau + \ldots$$

$$= W_C + W_D + W_Q + \ldots .$$

W_C, which is simply equal to

$$\phi(0) \int \rho_e(x, y', z') \, d\tau,$$

is seen to represent the Coulomb energy associated with the equivalent point charge. W_D represents the energy of the dipole moment of the charge distribution in the constant field

$$\left[\frac{\partial \phi(0)}{\partial x}, \frac{\partial \phi(0)}{\partial y'}, \frac{\partial \phi(0)}{\partial z'} \right].$$

When the charge distribution being considered is that of a nucleus $W_D = 0$, since the electrical dipole moment is equal to zero.

We now consider the third term W_Q. Because of the axial symmetry, ϕ depends on x and y' only through its dependence on the distance from the z'-axis. Hence in ϕ, x and y' appear only in the combination $\sqrt{(x^2 + y'^2)}$. It follows that

$$\frac{\partial^2 \phi(0)}{\partial x \, \partial y'} = \frac{\partial^2 \phi(0)}{\partial x \, \partial z'} = \frac{\partial^2 \phi(0)}{\partial y \, \partial z'} = 0.$$

Further, from symmetry,

$$\frac{\partial^2 \phi}{\partial x^2} = \frac{\partial^2 \phi}{\partial y'^2}.$$

Also, if we assume that ϕ arises from the atomic-electron Coulomb charge, and if we neglect the charge density of the orbital electrons within the nuclear volume, then ϕ satisfies Laplace's equation at the origin,

i.e. $\quad \dfrac{\partial^2 \phi(0)}{\partial x^2} + \dfrac{\partial^2 \phi(0)}{\partial y'^2} + \dfrac{\partial^2 \phi(0)}{\partial z'^2} = 0,$

and so $\quad \dfrac{\partial^2 \phi(0)}{\partial z'^2} = -2 \dfrac{\partial^2 \phi(0)}{\partial x^2} = -2 \dfrac{\partial^2 \phi(0)}{\partial y'^2}.$

Hence $W_Q = \frac{1}{2} \int \rho_e(x, y', z') \left[z'^2 \frac{\partial^2 \phi(0)}{\partial z'^2} - \frac{1}{2}(x^2 + y'^2) \frac{\partial^2 \phi(0)}{\partial z'^2} \right] d\tau$

$$= \frac{1}{4} \int \rho_e(x, y', z')(3z'^2 - r^2) \frac{\partial^2 \phi(0)}{\partial z'^2} \, d\tau,$$

where $r = \sqrt{(x^2 + y'^2 + z'^2)}$. Thus finally we have

$$W_Q = \tfrac{1}{4} q \int \rho_e(x, y', z')(3z'^2 - r^2) \, d\tau,$$

where q is the gradient of the electric field intensity in the direction Oz'. We now express the integral in terms of the coordinates (x, y, z), which refer to the body axes Ox, Oy, Oz, noting that $z' = z \cos \beta - y \sin \beta$. Therefore

$$W_Q = \tfrac{1}{4} q \int \rho_e(x, y, z)(3z^2 \cos^2 \beta + 3y^2 \sin^2 \beta - 6yz \sin \beta \cos \beta - r^2) \, d\tau.$$

From the axial symmetry of the charge distribution,

$$\int yz \, \rho_e(x, y, z) \, d\tau = 0.$$

Also from symmetry

$$\int y^2 \rho_e(x, y, z) \, d\tau = \int x^2 \rho_e(x, y, z) \, d\tau$$

$$= \tfrac{1}{2} \int (x^2 + y^2) \rho_e(x, y, z) \, d\tau$$

$$= \tfrac{1}{2} \int (r^2 - z^2) \rho_e(x, y, z) \, d\tau.$$

Therefore $W_Q = \tfrac{1}{4} q \int \rho_e(x, y, z) [3z^2 \cos^2 \beta + \tfrac{3}{2}(r^2 - z^2) \sin^2 \beta] \, d\tau$

$$= \tfrac{1}{8} q \left[\int \rho_e(x, y, z)(3z^2 - r^2) \, d\tau \right] (3 \cos^2 \beta - 1)$$

$$= \tfrac{1}{8} q e Q_0 (3 \cos^2 \beta - 1), \qquad\qquad 7.13$$

where Q_0 is the quadrupole moment of the charge distribution as defined in section 7.6.

Equation 7.13 gives the interaction energy of a nucleus having a quadrupole moment Q_0 when its symmetry axis lies at an angle β to the symmetry axis of an axially symmetrical electric field with a field gradient q at the site of the nucleus.

7.9 Quadrupole moments in quantum-mechanical systems

The nucleus has as its axis of averaged symmetry the axis of the total angular momentum I. We now consider the nucleus to be placed in an electric field of axial symmetry at a point in the field where the gradient along the field symmetry axis Oz' is q. The couple arising from the interaction of the quadrupole moment with the field gradient will cause I to precess about Oz'. The usual quantum-mechanical conditions will require that the angle between I and Oz' be such that the component of angular momentum parallel to Oz' is $M_I \hbar$, where M_I is integral if I is integral, M_I is half-integral if I is half-integral, and where $|M_I| \leqslant I$. Since the angle of precession β can never be zero, the value of W_Q can never attain the value it would have in the classical case for a charge distribution with quadrupole moment Q_0, where the angle β could be zero. In the quantum-mechanical case the minimum value of the angle β will be

$$\cos^{-1} \frac{I}{\sqrt{[I(I+1)]}},$$

and the interaction energy for this orientation is given by

$$W_Q = \tfrac{1}{8} q e Q_0 \left[3 \frac{I^2}{I(I+1)} - 1 \right]$$

$$= \tfrac{1}{8} q e Q_0 \left[\frac{2I-1}{I+1} \right]. \qquad \qquad \text{7.14}$$

We now introduce Q as the effective quadrupole moment, that is, the quadrupole moment which is observed experimentally. It will be less than the intrinsic quadrupole moment Q_0, because of the averaging of the charge distribution by the necessary precession of I. From 7.13, the maximum value of W_Q is

$$\tfrac{1}{8} q e Q (3 \cos^2 0° - 1) = \tfrac{1}{4} q e Q.$$

Comparing this expression with equation 7.14, we see that

$$Q = \frac{2I-1}{2(I+1)} Q_0.$$

We now consider the expressions for the interaction energies of the magnetic substates. The angle β is given by

$$\cos^{-1} \frac{M_I}{\sqrt{[I(I+1)]}}$$

and hence

$$(W_Q)_{M_I} = \tfrac{1}{8} q e Q_0 \left[\frac{3M_I^2}{I(I+1)} - 1 \right]$$

$$= \tfrac{1}{8} q e Q \frac{2(I+1)}{2I-1} \left[\frac{3M_I^2 - I(I+1)}{I(I+1)} \right]$$

$$= \tfrac{1}{4} q e Q \left[\frac{3M_I^2 - I(I+1)}{I(2I-1)} \right]. \qquad \text{7.15}$$

We now have to consider the two special cases $I = 0$ and $I = \frac{1}{2}$. Let us assume that we are dealing with a nucleus whose shape is that of an ellipsoid of revolution having an axis of symmetry Oz. When $I = 0$ there will be no rotation about the symmetry axis Oz. The lack of angular momentum about this axis enables the nucleus to lie with this axis orientated at random with respect to any specified direction. The arguments made above are then no longer valid for this case. The randomness of orientation of the nucleus will ensure its equivalence to a spherically symmetric charge distribution. Thus, although it has a finite intrinsic quadrupole moment Q_0, the effective quadrupole moment Q is zero. Turning to the case of $I = \frac{1}{2}$, we note that β must have the value $\cos^{-1}(1/\sqrt{3})$ and hence the nuclear orientation is always such that $(3 \cos^2\beta - 1) = 0$. Hence, as is clear from the above expression for Q in terms of Q_0, again the effective quadrupole moment is zero despite the nucleus having an intrinsic quadrupole moment. We therefore arrive at the important conclusion that the measured quadrupole moment can only be finite for $I \geqslant 1$. This is found to be in accord with experimental measurements.

7.10 Summary

The theoretical concept of magnetic dipole moment was developed to establish, on the basis of the single-particle hypothesis, the relationship between the single-particle quantum numbers and the nuclear dipole moment.

The static electric moments of the nucleus were defined, particular attention being paid to the electric quadrupole moment and its role in connection with the classical interaction energy of the nucleus with an electric field. The modifications to these results arising from a quantum-mechanical treatment were discussed.

Chapter 8
Nuclear Moments 2:
Experimental Measurements

8.1   **The determination of nuclear moments from optical spectrometry**

8.1.1   *Early history of optical spectroscopy*

As early as 1924, Pauli suggested that the hyperfine structure which had been observed in optical spectra, and which at the time lacked a satisfactory explanation, could possibly be arising from the effects of a *nuclear* magnetic moment. However, considerable advances in the theoretical understanding of the atomic system and its associated optical spectrum were necessary before this suggestion could be tested experimentally. We proceed briefly to summarize the interpretation of optical spectra with a view to outlining the role that the nuclear electric and magnetic moments play in this field of study so that it may be seen how information concerning the magnitude of these moments is to be obtained from spectral measurements. An understanding of the ideas and concepts of atomic physics is particularly relevant to our study of the nucleus since nuclear physics has borrowed heavily from the established ideas and concepts of atomic physics.

During the nineteenth century, extensive studies were made with optical spectrometers of the line spectra emitted and absorbed by the various chemical elements. Each element was found to have its own characteristic pattern of lines, a fact which was of the greatest practical importance for the identification of the elements in chemical and astronomical applications. The first step towards an understanding of the physical implications of the spectral lines was taken by Balmer in 1885. He found that if, instead of considering wavelength λ, whose measurement was in general the object of the experimental researches, one worked with the wave number k, then the nine known lines in the visible and ultraviolet spectrum of hydrogen were well fitted by the formula

$$k = \frac{1}{\lambda} = R \left[\frac{1}{2^2} - \frac{1}{m^2} \right],$$

with $m = 3, 4, 5, \ldots$, etc. with a suitable choice of the value for R, called the Rydberg constant. Now, since $\lambda = c/\nu$, its reciprocal k is proportional to the frequency ν. The reason for the success of Balmer's empirical formula became clear with the introduction of the quantum concept. On the quantum view the energy is emitted in the form of photons, each photon having an energy content given by $h\nu$. Thus the energy of the photon is proportional to k. Balmer's formula can then be interpreted as expressing the photon energy as the difference between a fixed energy and a set of discrete energy values. This led to the picture of the atom having a series of excited states, each of well-defined energy, the energies of these states forming the *terms* in the Balmer formula and the emitted photon carrying off the excess energy when a transition from a higher to a lower energy state occurs. On this view absorption finds an explanation in the transition from lower to higher energy states. A knowledge of the wavelengths of the optical spectral lines thus enables us to find very accurately the difference in energy content of the various excited states of the atom.

Bohr in 1913 introduced the assumption that in hydrogen stationary (i.e. non-radiating) states correspond to circular orbits of the electron of radius such that the angular momentum is $n\hbar$, where n is an integer. This led to energy values which corresponded with precision to the terms in Balmer's formula. n was referred to as the *quantum number*; its value determined the energy of the state.

8.1.2 Fine structure

When it became possible to examine the Balmer lines with higher resolution it was discovered that they were not in fact single but consisted of groups of lines called *multiplets*. Whereas the spacings between the lines given by the Balmer formula was of the order of tens of nanometres, the spacing between the members of a multiplet was of the order of hundredths of nanometres. This difference represents the justification for referring to the multiplets as constituting *fine structure*. Sommerfeld (1915) sought to explain fine structure by introducing elliptic orbits requiring a second quantum number n_θ for their description. In the non-relativistic approximation, and in the Coulomb field, the energy corresponding to an orbit is independent of n_θ for the orbit, being the same for the whole set of ellipses (including the circle), with the same value of n, which now became known as the *principal* quantum number. This degeneracy is removed when the relativistic mass dependence on velocity is introduced or when the field differs significantly from the Coulomb field; the energy of the system then depends slightly on n_θ.

These quantum-based ideas, developed to explain the single-electron hydrogen atom, had also a limited success when applied to the more complicated atoms of the alkali elements. In so far as the electronic configuration of these elements consisted of one electron orbiting outside one or more closed shells of electrons, the optical activity associated with this electron was expected to show hydrogen-like behaviour. However the field of the nucleus with its charge Ze screened by the $Z - 1$ electrons in the closed shells is not precisely Coulombic, even beyond the outer shell. Moreover, electrons on certain elliptic orbits would be expected to penetrate the shells. Thus the fine structure is expected to be more marked. Despite the complications certain progress was made in interpreting series of lines observed in terms of atomic energy levels. The Sommerfeld interpretation of fine structure was not however entirely satisfactory and in 1925 Goudsmit and Uhlenbeck proposed the idea of electron spin to justify the introduction of a further quantum number. This proposal also had the merit of resolving certain difficulties which at that time existed in interpreting the results of atomic-beam experiments.

In 1928 the theory was put on a much more satisfactory basis with Dirac's formulation of a relativistic wave equation for the electron, in which electron spin found a natural role. Dirac's theory led to a formula for fine structure in which there was a contribution from electron spin and also a contribution from the relativistic mass effect.

It is not however usually necessary to introduce the complications involved by having recourse to the Dirac formula. For many purposes it is sufficient to use the non-relativistic Schrödinger equation for a point electron in a field whose potential is appropriate to the atom under consideration, and to make separate allowance for the energy of interaction of the magnetic moment of the electron with the magnetic field arising from the relative motion of electron and charged nucleus. This interaction is the so-called *spin–orbit interaction.*

The solution of Schrödinger's equation then involves, as in the analysis of section 6.3, quantum numbers n, l, m, the eigenvalues being, as in the Bohr treatment, degenerate with respect to l and m. This degeneracy is removed when the field is not strictly Coulombic or when allowance is made for the relativistic mass effect. The introduction of electron spin s leads to the modification of the quantum numbers. These, in the presence of spin–orbit coupling, now become n, l, j, m_j, where $j = l \pm s$ and m_j is the projection of j on a specified axis.

The probability of transitions occurring between two particular states, which governs the intensity of the associated spectral line, is strongly dependent on the quantum numbers associated with the initial and final states. The transitions are most likely, and hence lead to observable line intensities, when certain *selection rules* are obeyed. For this reason it is very important to be able to label states of the atomic system with the appropriate quantum numbers.

In the case of one-electron atoms (e.g. hydrogen, singly ionized helium) the quantum numbers to be associated with the states are those of the single electron. In the case of the alkali elements, the optical activity is mostly associated with the valence electron. The closed shells having zero total orbital angular momentum and zero total intrinsic electron spin, the atomic states again can be satisfactorily labelled with the quantum numbers of the single valence electron.

In the case of other atoms where more than one electron is involved outside a closed shell, the quantum numbers of the atomic state have to be constructed from those of the electrons involved. Normally this is carried through by combining vectorially the individual orbital angular momenta to form a resultant orbital angular momentum **L** and the spin angular momenta to form a combined spin angular momentum **S**. A total angular momentum **J** is then formed by combining **L** and **S** vectorially. Associated with the quantum number J is the magnetic quantum number M arising from the projection of **J** on a specified axis.

The selection rule which is found normally to operate is that, between initial and final atomic states, $\Delta J = \pm 1$ or 0, with the usual proviso that a transition from a state with $J = 0$ to a second state with $J = 0$ is forbidden. It is usual for there to be no change in S so that $\Delta L = \pm 1$ or 0.

Analogous with the energy dependence on orientation of a magnetic dipole in a magnetic field in classical physics, there is a dependence of the atomic energy on the orientation of **S** with respect to **L**. Thus we can have states with the same values of L and S, but different J-values, which differ from each other slightly in energy content.

As an example of the application of these ideas let us consider the neutral atom of mercury. We consider its complement of eighty electrons to be arranged as follows. The first five shells are complete, i.e. mercury has a core of similar electronic configuration to the atom of xenon ($Z = 54$). In the sixth shell there are complete subshells corresponding to fourteen electrons with $n = 4$, $l = 3$, together with all the possible values of j and m_j, ten electrons with $n = 5$, $l = 2$ and two electrons with $n = 6$, $l = 0$. We can now consider excited states of the atom in which one of the latter two s-electrons is displaced into an orbit of higher energy, this being the sole change in the configuration. The electrons in the completed shells and subshells have zero total angular momentum and zero total spin. Hence to find L and S we need only consider the last two electrons. Of these, the undisturbed member of the pair has $l = 0$. Thus the value of L will be equal to the l-value of the disturbed electron, while S may be 1 or 0 depending on the parallelism or antiparallelism of the electron spins. By promoting this single electron to an orbit with $L = 1$ the ground-state configuration, which we denote by S_0 (the conventional symbol for $L = 0$, $J = 0$), is changed into one of the following three configurations:

P_0 ($L = 1$, $J = 0$, arising from $S = 1$, **L** and **S** antiparallel),

P_1 ($L = 1$, $J = 1$, arising from $S = 1$, **L** and **S** at $120°$) and

P_2 ($L = 1$, $J = 2$, arising from $S = 1$, **L** and **S** parallel).

Similar considerations show that if the second electron is raised into an orbit with $l = 2$ we can then have configurations:

D_1 ($L = 2$, $J = 1$, $S = 1$),

D_2 ($L = 2$, $J = 2$, $S = 1$) and

D_3 ($L = 2$, $J = 3$, $S = 1$).

These groups of P- and D-states are said to constitute *triplets*, each having a *multiplicity* of three. Transitions between the components of the D-triplet and the components of the P-triplet take place, the probabilities of these transitions being governed by the selection rules stated above. In Figure 37 the levels corresponding to these states are illustrated and the wavelengths of the emitted quanta indicated to show the wavelength differences involved in the structure arising from the spin–orbit interaction.

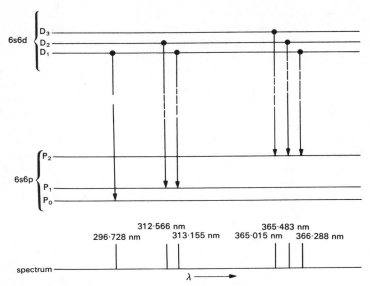

Figure 37 Interpretation of fine structure in optical spectrum of neutral mercury atom

8.1.3 *Hyperfine structure*

When the developments of spectroscopy made it possible to observe with still higher resolution, structure even finer than the fine structure we discussed above was discovered. For example a 'line' of wavelength 359·6 nm in the spectrum of bismuth was found to contain six components within a wavelength range of 0·03 nm (equivalent to a range of wave numbers of 230 m^{-1}). Structure of this order of fineness is referred to as *hyperfine structure*.

In some cases hyperfine structure is due to the existence in the sample of more than one stable isotope of the element being investigated. When the sample consists of an isotopic mixture, there appear to be two distinct reasons for slight differences occurring in the spectral lines emitted by the different isotopes, in spite of the fact that they must be assumed to have identical electron configurations. Firstly, the different nuclear masses of the isotopes lead to different reduced masses (see section 3.6) for the nucleus–electron system. The different reduced masses reflect the different relationships between the centre-of-mass and laboratory coordinate systems, and slightly affect the energy. The differences between the spectrum of hydrogen ($Z = 1, A = 1$) and deuterium ($Z = 1, A = 2$) is well explained on this basis. In fact this was a central consideration in the argument made by Urey in 1932 for the existence of a heavy isotope of hydrogen. However, in the case of the heavier elements this explanation is not satisfactory and the isotope shift may in fact be in the opposite direction to that predicted by making allowance for the change in reduced mass. In these cases the dominant cause is believed to be a dependence of the nuclear radius, and consequently of the Coulomb field close to the nucleus, on the number of neutrons in the nucleus.

However, hyperfine structure is not solely isotope structure. This became clear when hyperfine structure was discovered in an element like bismuth which is known to be mono-isotopic. Thus a further explanation for hyperfine structure had to be sought. We recollect that in the case of fine structure the introduction of a new quantum number s, the electron spin, and of a new interaction, namely spin–orbit interaction, provided a satisfactory theoretical basis for its discussion. So in the case of hyperfine structure we resort to the same expedients. We recall Pauli's suggestion concerning the nuclear moments and we take the new quantum number to be I, the nuclear spin. The new interaction to be considered is the interaction between the magnetic and electric moments of the nucleus, which are related to its spin and its charge distribution, and the electric and magnetic field at the site of the nucleus due to the orbital atomic electrons.

Just as we combined $\mathbf{L}$ and $\mathbf{S}$ to form $\mathbf{J}$, so we now combine $\mathbf{J}$ and $\mathbf{I}$ to form $\mathbf{F}$ which will be taken to represent the angular momentum of the whole atom, including the contribution from the intrinsic nuclear spin. Again, analogous to the previous case, the energy of the system depends slightly on the orientation of I with respect to J. Thus states with the same J and I quantum numbers but differing in F quantum number can have slightly different energies. We note that F can take the values $J + I, J + I - 1, J + I - 2, \ldots, |J - I|$, giving $2J + 1$ different values providing $J \leqslant I$, and $2I + 1$ different values if $J \geqslant I$.

Assuming J to be known from the interpretation of fine structure, then the number of hyperfine structure components associated with the various states provides important information about the value of I. If $J = 0$, there is of course only one possible value of F, irrespective of the value of I, and hence all states with $J = 0$ are expected to be singlets. If $I = \frac{1}{2}$ then, apart from states with $J = 0$, all states will have two hyperfine components. If $I > \frac{1}{2}$ then there will be $2J + 1$ or $2I + 1$ components, depending on whether $J \leqslant I$ or $J > I$. Transitions between components of two hypermultiplets will of course be governed by a selection rule which is found to be $\Delta F = \pm 1$ or 0, with $F = 0 \rightarrow F = 0$ forbidden. We now take the particular example of a hypermultiplet namely that of the 377·7 nm ($k = 2647 \cdot 8$ mm^{-1}) line in thallium. This is believed to be a $P_{\frac{1}{2}} \rightarrow S_{\frac{1}{2}}$ transition. If we tentatively assign a value of $\frac{1}{2}$ to the spin of the thallium nucleus, then in the case of the upper-state J-value being $\frac{1}{2}$, we can have $F = 1$ or $F = 0$. Again, J being $\frac{1}{2}$ for the lower state, F can be 1 or 0 for this state. The energy levels and transitions are illustrated in Figure 38 and these provide a satisfactory explanation for the pattern of lines observed.

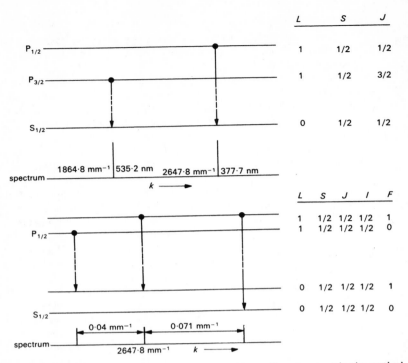

L	S	J
1	1/2	1/2
1	1/2	3/2
0	1/2	1/2

1864·8 mm⁻¹ | 535·2 nm   2647·8 mm⁻¹ | 377·7 nm

L	S	J	I	F
1	1/2	1/2	1/2	1
1	1/2	1/2	1/2	0
0	1/2	1/2	1/2	1
0	1/2	1/2	1/2	0

0·04 mm⁻¹   0·071 mm⁻¹

2647·8 mm⁻¹   k ⟶

Figure 38 Interpretation of fine structure and hyperfine structure in the optical spectrum of thallium

Consider the further more complicated example arising in the case of singly ionized praseodymium. The J-values of the states are believed to be large. We take them to be 7 and 8, although the argument is unaffected providing they are greater than I, which we tentatively take to be $\frac{5}{2}$. Then as illustrated in Figure 39 each state has $2I + 1$, i.e. six, components. The above selection rule then predicts the line pattern shown and this is in good agreement with observations. Thus the assignment of the value $\frac{5}{2}$ to I is confirmed. Any other value of I (subject always to $J > I$) can be shown to predict a different appearance for the hypermultiplet.

We thus have, always providing some information is available concerning the J-values, a very powerful method for determining I from the line pattern alone. We now turn to the other information available which we have not so far used, namely the spacing between the lines in the spectrum. This spacing is of course related to the spacing between the energy levels and in turn this is related to the interaction energy associated with the orientation of the nucleus.

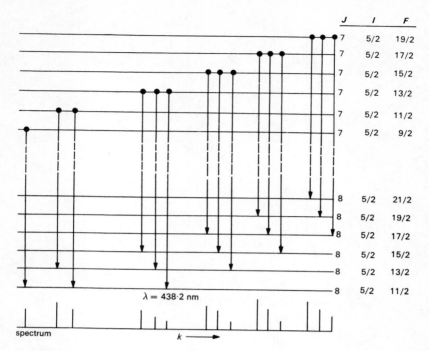

J	I	F
7	5/2	19/2
7	5/2	17/2
7	5/2	15/2
7	5/2	13/2
7	5/2	11/2
7	5/2	9/2
8	5/2	21/2
8	5/2	19/2
8	5/2	17/2
8	5/2	15/2
8	5/2	13/2
8	5/2	11/2

$\lambda = 438 \cdot 2$ nm

spectrum $k \longrightarrow$

Figure 39 Interpretation of hyperfine structure of an optical line in the spectrum of Pr$^+$

8.1.4   *The determination of nuclear magnetic moments*

Consider first the effect of the nuclear magnetic moment. As discussed in section 7.3, this quantity is related to the spin angular momentum and we take the magnetic moment to be μ_I parallel to I, the absolute angular momentum. We note that μ_I, like I, is not observable and that the maximum observable time-averaged magnetic moment is given by

$$\mu_I = \mu_I \frac{I}{\sqrt{[I(I+1)]}}.$$

This effective magnetic moment will interact with the magnetic field produced at the nucleus by the orbital atomic electrons. We assume this field to be H_J, antiparallel to J. (The antiparallelism is taken because the negative charge of the electron leads to this relative orientation for a single-electron atom. The quantities may of course be parallel for multi-electron atoms.) Since the vector J must always precess about a specified direction, the effective magnetic field will always be less than H_J. We denote the effective magnetic field by H_J, where

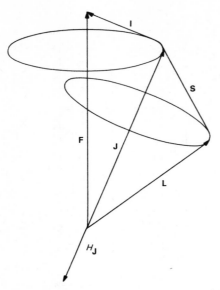

Figure 40 Vector diagram of coupling of L, S and I

$$H_J = H_J \frac{J}{\sqrt{[J(J+1)]}}.$$

We assume that the vectors $\mathbf{I}$ and $\mathbf{J}$ couple together to form a resultant vector $\mathbf{F}$ as shown in Figure 40. The contribution to the total energy of the system arising from the interaction of the nuclear magnetic dipole moment with the atomic magnetic field is then

$$W_D = \mu_I H_J \cos(\mathbf{I}, \mathbf{J}).$$

Now from Figure 40 we see that

$$\cos(\mathbf{I}, \mathbf{J}) = \frac{|\mathbf{F}|^2 - |\mathbf{J}|^2 - |\mathbf{I}|^2}{2|\mathbf{I}||\mathbf{J}|}.$$

Hence $W_D = \mu_I H_J \dfrac{F(F+1) - J(J+1) - I(I+1)}{2IJ}.$ 8.1

It is customary to write this expression as

$$W_D = \frac{AC}{2},$$

where $A = \dfrac{\mu_I H_J}{IJ}$ and $C = F(F+1) - J(J+1) - I(I+1).$

We now take I and J as fixed quantities and consider F as a variable quantity subject to the condition that its allowed values are $I+J, I+J-1, \ldots, |I-J|.$

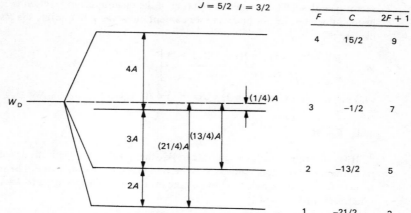

Figure 41 Interval rule for magnetic-dipole interaction. μ_I is taken to have a positive value and H_J is assumed antiparallel to J. Note intervals $4A$, $3A$, $2A$ proportional to upper F-value

If A is positive, then the higher the value of F the greater the energy associated with the state. By substituting into equation 8.1 we see that the energy difference between a state having a quantum number F and one having a quantum number $F - 1$ will be AF. This constitutes the *interval rule* for the magnetic dipole interaction. Figure 41 shows the intervals involved in the case where $J = \frac{5}{2}$, $I = \frac{3}{2}$, μ_I is positive and H_J is antiparallel to J.

In the same diagram the displacement of each component with respect to a base line corresponding to $W_D = 0$ is shown. These displacements are given by $\frac{1}{2} AC$ for the different F-values. We note that if, corresponding to the possible orientations for each F-value with respect to a specified spatial direction, a statistical weight $2F + 1$ is assigned, and the 'centroid' of the multiplet calculated from

$$\sum (2F + 1)(W_D)_F,$$

then the energy corresponding to the centroid is found to be $W_D = 0$. This means that the multiplet components have a centroid corresponding to the level as it would be determined in an experiment with insufficient resolution to separate the components.

8.1.5 The determination of nuclear electric quadrupole moments

We now consider the interaction between the electric quadrupole moment of the nucleus and the gradient of the electric field at the nucleus. The integration energy is the scalar product of two tensors of degree two, namely the quadrupole moment and the field gradient. This is to be compared with the case of the magnetic dipole interaction, in which the tensors were of degree one, i.e. were

133 The determination of nuclear moments from optical spectrometry

vector quantities. The detailed calculation of the quadrupole interaction is beyond the scope of this book and we content ourselves with quoting the results, namely that

$$W_Q = \frac{B}{4}\left[\frac{\frac{3}{2}C(C+1) - 2I(I+1)J(J+1)}{I(2I-1)J(2J-1)}\right],$$

where as before $C = F(F+1) - J(J+1) - I(I+1)$,

and $B = eQ\dfrac{\partial^2\phi(0)}{\partial z^2}$.

If W_Q is comparable in magnitude to W_D there will arise significant deviations from the magnetic dipole interval rule derived in section 8.1.4. This can be seen by considering the case of the hyperfine multiplet corresponding to $J = 1, I = \frac{3}{2}$ illustrated in Figure 42.

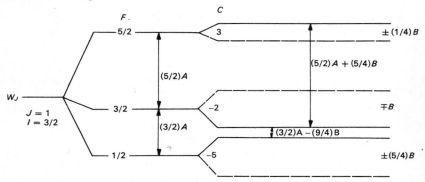

Figure 42 Effect of finite quadrupole moment on the hyperfine structure levels. Full lines are for $B > 0$, dotted lines for $B < 0$

The first evidence of the existence of a nuclear electric quadrupole moment consisted of these measurable deviations from the magnetic dipole interval rule.

8.2 Limitations to the accuracy of optical determinations of nuclear moments

Accurate optical measurements of hyperfine splitting in conjunction with the above theoretical interpretation enables values of A and B to be determined in many cases. In order to derive μ and Q from these quantities, however, one requires the values of magnetic field gradient and electric field gradient at the nucleus. These fields have to be calculated from the quantum mechanics of the electron orbitals. Apart from the most elementary cases, the accuracy of the calculated values, particularly those of electric field gradients, are open to some doubt. However, to give an impression of the values of fields involved, we quote the result of the calculations for caesium. This atom has one electron in excess of the configuration of xenon. In the ground state the valence electron is in an s-state. While it has no orbital angular momentum, nevertheless by virtue of its intrinsic magnetic moment of one magneton, it produces at the nucleus a magnetic field of $2 \cdot 1 \times 10^6$ G. We note that an electron at a distance of 10^{-10} m produces a magnetic field of approximately 10^4 G; hence the effectiveness of the s-electron is due to its greater proximity to the nucleus. In fact it produces a greater magnetic field than an electron in a p-orbit, and it is calculated that the value of $H(0)$ for the ground state is about ten times that for the excited states corresponding to p-electrons. The s-electron in the ground state has spherically symmetric charge distribution and does not therefore give rise to an electric field gradient at the nucleus. If however we take the excited states, then the lowest-energy p-electron gives rise to an electric field gradient $\partial^2\phi(0)/\partial z^2$ of $1 \cdot 1 \times 10^{22}$ V m^{-2}. Note that a charge distribution equivalent to one electronic charge at a distance of 10^{-10} m from the nucleus would produce a gradient of about 3×10^{21} V m^{-2}.

We therefore see that, in principle, values of μ and Q can be obtained from the measured values of A and B using the calculated field values. In the case of the magnetic moments the results are in fairly good agreement with results obtained by other methods, to be described below. However in the case of the electric quadrupole moments the situation is much less satisfactory. This is so for two reasons. Firstly because the interaction energy itself is smaller and secondly because there is more doubt about the effective field gradient at the nucleus than there is about the magnetic field.

Where information concerning the hyperfine splitting is available for two isotopes of the same element, then the ratio of the μ-values or Q-values is much more reliable because the ratio does not depend on the field at the nucleus, which is assumed to be the same for the same electron configurations. From the ratios one can learn how the nuclear magnetic moments are being affected by the addition of neutrons to a particular nucleus. However, even in the case of isotopes there is the possibility that the electron orbitals may be affected by the nuclear quadrupole moment and might therefore not be the same for both isotopes.

The investigation of hyperfine structure calls for the highest available resolution because the very small energy difference between two components of a multiplet is being measured as the difference between two quanta each of very much higher energy. There are now available techniques whereby transitions between components in the multiplet may be directly investigated and as a consequence much greater accuracy can be achieved by these newer methods in the measurement of A and B. These newer techniques in general involve applying an external magnetic field to the atomic system. We now proceed to consider the consequences of the application of such an external field.

8.3   Behaviour of atomic system in a magnetic field

We recall that the system consists of a set of orbital electrons having a resultant angular momentum $\mathbf{J}$. The electrons will also in general have a resultant magnetic moment μ_J taken antiparallel to $\mathbf{J}$ and of magnitude of the order of the Bohr magneton ($9 \cdot 2731 \times 10^{-28}$ J G^{-1}). At the centre of the system we have a nucleus of angular momentum I and magnetic moment μ_I parallel to I of magnitude of the order of a nuclear magneton ($5 \cdot 05038 \times 10^{-31}$ J G^{-1}). We have already discussed how J and I couple to form a resultant F (see Figure 40, p. 132).

If now this system is immersed in a weak external magnetic field H, F will precess round H, the angle of the cone on which it lies being such that the angular momentum parallel to H is $M_F \hbar$, M_F as usual being integral or zero and $|M_F| \leqslant F$. From the vector diagram of Figure 40 it can be seen that the time-averaged value of magnetic moment parallel to F, which we denote by μ_F, will be given by

$$\mu_F = \mu_I \cos(I, F) - \mu_J \cos(F, J).$$

Thus
$$\mu_F = \mu_I \left[\frac{|F|^2 + |I|^2 - |J|^2}{2|I| \, |F|} \right] - \mu_J \left[\frac{|F|^2 + |J|^2 - |I|^2}{2|F| \, |J|} \right].$$

The interaction energy of the resultant magnetic dipole moment of the atomic system with the external field will be given by

$$W_H = - \mu_F H \cos(F, H) = -\mu_F H \frac{M_F \hbar}{|F|}$$

$$= - \frac{M_F \hbar H}{2|F|^2} \left[\frac{\mu_I}{|I|} (|F|^2 + |I|^2 - |J|^2) - \frac{\mu_J}{|J|} (|F|^2 + |J|^2 - |I|^2) \right].$$

This energy increment has to be added to the natural hyperfine energy $\frac{1}{2} A C$ of section 8.1.4.

We now illustrate these general results by considering a particular case. Take $J = \frac{1}{2}, I = \frac{3}{2}$; then $F = 2$ or 1. Taking $F = 2$ and neglecting μ_I compared to μ_J, we have

$$(W_H)_{F=2} = \frac{M_F \hbar H}{4} \frac{\mu_J}{|J|}, \qquad M_F = 2, 1, 0, -1, -2.$$

For $F = 1$ we have

$$(W_H)_{F=1} = - \frac{M_F \hbar H}{4} \frac{\mu_J}{|J|}, \qquad M_F = 1, 0, -1.$$

In Figure 43 is illustrated schematically the variation of energy as the external field H increases in value.

As H increases, the angular velocity of precession of F about H increases until the simple vector picture ceases to be valid. At very high field strengths F is precessing as fast about H as I and J are precessing about F. When this stage is

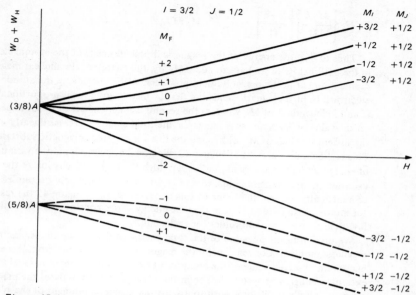

Figure 43 Variation of energy of atomic system as the externally applied field H is increased

reached $\mathbf{J}$ and $\mathbf{I}$ become decoupled and each then precesses independently about $\mathbf{H}$ with magnetic quantum numbers M_J and M_I. We then have W_H given by

$$\frac{\mu_{\mathbf{J}} M_J \hbar}{|\mathbf{J}|} H,$$

together with a term

$$\frac{\mu_{\mathbf{I}} M_I \hbar}{|\mathbf{I}|} H,$$

which we neglect. However, since the field at the nucleus arising from the orbital electrons is large compared to the external field, we have to allow for the interaction of the nuclear dipole moment with this internal field. The angle between $\mathbf{I}$ and $\mathbf{J}$ is not now constant as it was in the discussion in section 8.1.4. However the precession of $\mathbf{J}$ is fast enough now for us to assume that the effective field at the nucleus is

$$H_{\mathbf{J}} \frac{M_J \hbar}{|\mathbf{J}|},$$

parallel to the direction of the external field H. We note that $H_{\mathbf{J}}$ will be antiparallel to $M_J \hbar$. The interaction energy of $\mu_{\mathbf{I}}$ with this atomic field will then be given by

$$\mu_I H_J \left[\frac{M_J \hbar}{|J|} \right] \left[\frac{M_I \hbar}{|I|} \right] = \frac{\mu_I H_J}{IJ} M_I M_J.$$

Thus to W_H, the energy of the magnetic dipole moment of the atomic electrons in the external field, we have to add $AM_I M_J$, the energy of the nuclear magnetic dipole moment in the magnetic field produced at the nucleus by the atomic electrons. In high magnetic fields the first of these two terms is proportional to H and independent of M_I, whereas the second is independent of H but depends on the value of M_I. Thus as H increases, the total energy increases linearly with the different values of M_I, displacing the line for different magnetic substates. This is shown on the right of Figure 43. We now have considered the variation of energy with H in low fields and in very high fields. The discussion of the variation in intermediate fields, when I and J are not completely decoupled, is of a difficulty beyond the scope of this book. We show in Figure 43 the results for intermediate fields and note that the energy varies continuously from the low-field to the high-field values for each state.

We turn from the question of the total energy of the system to discuss the magnetic moment. Conventionally the magnetic moment of the atom is the maximum observed value of the component of magnetic moment parallel to an external field **H**. The system will be in this state for one set only of the magnetic quantum numbers. We now wish to discuss the magnetic moment of the system for any set of magnetic quantum numbers and find it convenient to introduce the term *effective magnetic moment*, denoted by μ_{eff}, for the component of magnetic moment parallel to H in the general case. The potential energy of the system can then be written

$$W_H = -\mu_{eff} H + \text{ a constant independent of } H.$$

It follows that

$$\mu_{eff} = -\frac{\partial W_H}{\partial H}.$$

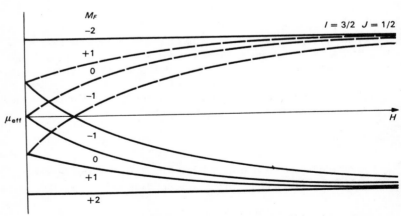

Figure 44 Variation of $\mu_{eff} = -\partial W_H/\partial H$, with increasing H for the system of Figure 43

We can thus derive μ_{eff} from the gradient of the curves in Figure 43. In low fields these values can be confirmed by taking the components parallel to **H** of $\mu_{\mathbf{F}}$ as given by equation 8.2. For high fields

$$\mu_{\text{eff}} = -\frac{M_J \hbar}{|J|}\mu_{\mathbf{J}}, \qquad M_J = \pm\tfrac{1}{2},$$

assuming the component of $\mu_{\mathbf{I}}$ to be negligible. In Figure 44 μ_{eff} is plotted. It is to be noted that μ_{eff} is zero not only in the obvious case of low fields with $M_F = 0$ but also in intermediate fields for certain other magnetic substates.

It is very important to note that the nucleus influences the behaviour of the atom, not only through its contribution to the energy of the system by virtue of the interaction of its dipole moment with the internal magnetic field, but also through the large contribution which its angular momentum makes to the total angular moment of the system. This is so despite the fact that its magnetic moment is three orders of magnitude smaller than the atomic magnetic moment.

8.4   Behaviour of atomic system in inhomogeneous magnetic field

A basic feature of the atomic-beam experiments which we wish to review is the use of inhomogeneous fields to deflect the beam. We proceed to discuss this deflection in relation to the effective magnetic moment of the atom and to the gradient of the external magnetic field.

Consider the simple case of a classical magnetic dipole, moment μ, lying with the moment parallel to a magnetic field H along the z-axis. If the field H is homogeneous there is no net force acting on the dipole. If, however, the field is inhomogeneous and there is a finite gradient at the origin denoted by $\partial H(0)/\partial z$, then from elementary magnetism we know that there is a resultant force

$$\mu\,\frac{\partial H(0)}{\partial z}$$

acting along the z-axis. The potential energy of the system with the dipole in the given orientation we can denote by $W_H = -\mu H$, measured with respect to a standard position with the dipole perpendicular to the z-axis. In this magnetic context W_H plays the role that electrostatic potential plays in an electrostatic field and just as the electric field strength is given by the negative gradient of the potential, so the force on the magnetic dipole is given by $-\partial W_H/\partial z$. But

$$-\frac{\partial W_H}{\partial z} = -\frac{\partial W_H}{\partial H}\cdot\frac{\partial H}{\partial z} = \mu\,\frac{\partial H}{\partial z},$$

confirming the result which is otherwise obvious in the elementary case.

For a more complicated system the deflecting force is still equal to

$$-\frac{\partial W_H}{\partial z} = -\frac{\partial W_H}{\partial H}\cdot\frac{\partial H}{\partial z} = \mu_{\text{eff}}\,\frac{\partial H}{\partial z}.$$

8.5 Measurement of nuclear spin and nuclear, magnetic and electric moments

The techniques, other than optical spectroscopy, which provide information concerning nuclear spin and magnetic moments can be said in general to fall into two groups; those employing atomic or molecular beams, in which there is no communication of energy between the participating atoms or molecules, and those involving liquid or solid samples where energy can pass from one atom or one molecule in the system to others in the sample.

8.5.1 *Beam-deflection measurements of nuclear dipole moments*

These measurements involve sending the beam of atoms or molecules along, let us say, the x-direction through an evacuated region in which there is a magnetic field at right angles, say in the z-direction, as in Figure 45. This field, which we denote by H_z, is designed to have a high value of the gradient $\partial H_z/\partial z$. Such a field can be achieved by the use of two electrical conductors carrying oppositely directed electric current. If no ferromagnetic materials are used then H_z and $\partial H_z/\partial z$ can be calculated with accuracy. If, however, higher field values are required these can be achieved from practical current densities by introducing ferromagnetic poles shaped to follow the magnetic equipotentials. In this case the field gradients have to be measured because of uncertainty about the behaviour of the ferromagnetic material and considerable practical difficulties may be entailed.

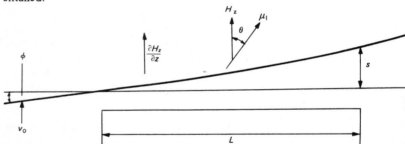

Figure 45 Trajectory of particle with magnetic moment through a magnetic field of constant gradient

Let us now consider classically the trajectory through such a field of an atomic system which has angular momentum I and magnetic moment μ_I parallel to the spin axis. If the dipole axis lies at an angle θ to the field direction then it will experience a couple of moment $H_z \mu_I \sin \theta$ about an axis normal to the plane containing H_z and μ_I. The existence of angular momentum however ensures that this couple does not simply align μ_I and H_z. The necessary increase of angular momentum created by the couple is achieved by rotation of the (H_z, μ_I) plane about the z-axis. The angular velocity of this rotation, called the *Larmor precession*, we denote by ω_L. It follows from the laws of classical dynamics that the rate of increase of angular momentum, $|I|\omega_L \sin \theta$, is equal to the moment of the applied couple, $\mu_I H_z \sin \theta$. Hence

$$\omega_L = \frac{\mu_I}{|I|} H_z.$$

It is important to note that ω_L is independent of θ, the orientation of the dipole with respect to the field.

The dipole experiences, in addition to the couple, a net translational force in the z-direction, as discussed in section 8.4. In this case the net force is

$$\mu_I \cos\theta \ \frac{\partial H_z}{\partial z},$$

in the z-direction. (The gradient $\partial H_z/\partial z$ must, from the field equations for a static field, be associated with a gradient in the x- and/or y-directions. Strictly speaking there should therefore be one or more additional terms in the expression for the force. However, any such additional terms reverse in direction as the dipole precesses and their total effect is averaged out over one cycle of the Larmor precession.) The acceleration f along the z-direction will then be given by

$$\frac{\mu_I \cos\theta}{M} \frac{\partial H_z}{\partial z}.$$

If the atom enters the system on the x-axis with a velocity v_0 and at an angle ϕ to the x-axis, and travels a distance L through the field, then it will spend a time $L/v_0 \cos\phi$ in the field and, from elementary kinematics, may be shown to travel a distance

$$s = L\tan\phi + \frac{1}{2} \frac{fL^2}{v_0^2 \cos^2\phi}$$

from the x-axis before it leaves the field. We can alternatively express the deflection as

$$s = L\tan\phi + \frac{1}{4} \frac{\mu_I(\cos\theta)L^2}{T\cos^2\phi} \frac{\partial H_z}{\partial z},$$

where T is the kinetic energy of the atom as it enters the field.

It is interesting to calculate the deflection that can be expected in a practical situation. With $\partial H_z/\partial z = 10^7$ G m^{-1}, $T = 1 \cdot 4 \times 10^{-20}$ J (which corresponds to the atoms being in thermal equilibrium at 1000 K), $\mu_I \cos\theta$ of the order of a Bohr magneton (9×10^{-28} J G^{-1}) and $L = 0 \cdot 5$ m, an atom entering along the axis will be deflected 45 mm. If, on the other hand, the magnetic dipole moment were of the order of nuclear rather than Bohr magnetons, then the deflection would be about $2 \cdot 5 \times 10^{-2}$ mm.

The first experiments involving this technique were those of Stern and Gerlach (1922) which first established spatial quantization and played an important historical role in the development of quantum theory. In the Stern and Gerlach experiments a beam of silver atoms was used and it was discovered that the deflected atoms fell into two separated groups. This was interpreted as arising from the spatial quantization of atomic dipoles, which were capable of having only two magnetic quantum numbers. This meant that J had to be equal to $\frac{1}{2}$. A certain spread about the discrete s-values is to be expected because of the variation of v_0 in the beam emitted from the oven which acted as the source of neutral silver atoms. In an experiment of this kind, should the nucleus have spin and magnetic moment, then in principle this should alter the observed deflection. The values of H_z normally used are strong enough completely to decouple $\mathbf{J}$ and $\mathbf{I}$ and consequently, as discussed in section 8.3, the magnetic moment is only altered by an amount of the order of a nuclear magneton. From the numerical example cited above it is clear that the associated change in deflection could not be detected because of the effect of the velocity variation in the atoms in the beam. If however the nucleus had spin $\frac{3}{2}$, then at some reduced field the effective atomic dipole moment, again as discussed in section 8.3, should vanish and there should be no deflection of the atom. Thus a collimator defining a very small solid angle with respect to the beam source, set up beyond the deflecting magnet, will transmit a beam which, as H_z is decreased from high values, will at some critical intermediate value show a marked rise in intensity. This effect is the more marked since the effective dipole moment vanishes at the same H_z value for atoms of all velocities. Quantum theory has then to be invoked to relate the critical field to the nuclear moment. This method of measuring nuclear magnetic dipole moments is known as the *zero-moment technique*.

In another application a beam of molecules, having no magnetic moment arising from the orbital electrons, is used. If in this case any deflection is observed then it must be due either to the nuclear moment, to a moment associated with rotation of the molecule or to diatomic moments induced in the electron system by the applied field. All of these moments can be of the same order and the separation of the effects is extremely complicated. Nevertheless it was possible by this method (Frisch and Stern, 1933) to make the first measurement of the proton magnetic moment using a beam of hydrogen molecules.

8.5.2   *Magnetic resonance beam methods*

A new order of accuracy in the measurement of nuclear magnetic dipole moments became possible with the application by Rabi and his associates (in 1939) of

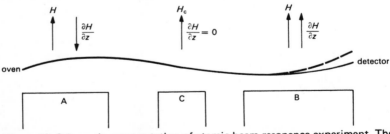

Figure 46 Schematic representation of atomic-beam resonance experiment. The resonant disturbance of the beam is arranged to take place in magnet C in which there is no gradient of magnetic field

resonance techniques to atomic and molecular beams. Schematically the apparatus is shown in Figure 46. The magnet A is a deflecting magnet of the inhomogeneous type described in section 8.5.1. Magnet B is a second inhomogeneous magnet arranged to have H_z in the same direction but the gradient $\partial H_z/\partial z$ reversed as compared to A. C is a magnet with a uniform field H_C. When H_C is zero, magnet B can be adjusted to compensate for the effect of magnet A and to bring the original beam on to the detector situated on the axis of the apparatus. This restoration of the beam by the magnet B is independent of the velocity of the atoms or molecules in the beam. When the field H_C is increased from zero the particles in the beam by virtue of their magnetic dipole moment will have a Larmor precession but will undergo no deflection because magnet C has a homogeneous field. The compensating effect of B is dependent on the atom or molecule maintaining its original magnetic state, i.e. the orientation θ has to be the same in B as it was in A. The Larmor precession in the magnet C does not of itself change the magnetic state of the atom or molecule. If, however, we arrange for a magnetic field H_0 to be applied perpendicular to H_C, and to rotate about H_C, there will be a tendency for precession to take place about H_0 as well as about H_C. This means that there will be a tendency for the orientation θ to alter. If H_0 rotates with an angular velocity different from that of the Larmor precession then the disturbance to θ will be sometimes tending to increase its value, at other times tending to reduce it. The net effect will average to zero. Should, however, H_0 rotate with an angular velocity exactly equal to the Larmor angular velocity then there will be a disturbance which will act steadily to alter θ in one direction. The simplest way to achieve a rotating magnetic field is to feed high-frequency electric current to a fixed coil having its plane arranged to contain the direction of the magnetic field H_C. This will produce a field $H \sin \omega t$ in a fixed direction perpendicular to H_C. Formally this is equivalent to the superposition of two fields, each of strength $\frac{1}{2}H$, rotating in opposite senses with angular velocity ω in a plane perpendicular to H_C. If $\omega = \omega_L$, one of these fields will produce on average no effect on the precessing system, as it is rotating in the direction opposed to that of the precession. The other field however will rotate in the opposite direction and represent the achievement of the resonant disturbance discussed above. The energy of the quantum associated with the high-frequency field will be

$$h\nu = \frac{h}{2\pi} \omega_L = \frac{\mu}{I} H_C.$$

The energy of the magnetic substate corresponding to magnetic quantum number M_I is given by

$$\frac{M_I}{I} \mu H_C.$$

Thus, as we would expect, the quantum of the resonant radio-frequency field has exactly the value required for a transition from a state with magnetic quantum number M_I to one with magnetic quantum number $M_I \pm 1$. We must take both algebraic signs into account because, just as in the classical description the disturbing torque may either decrease or increase the orientational energy, on a quantum treatment the perturbing field may stimulate quantum emission as well as lead to quantum absorption. In either of these events the magnetic state of the atom or molecule has been changed, and for that particular particle magnet B now no longer exactly compensates for the effect of magnet A. Thus there will be a drop in intensity of the beam recorded at the detector. In practice the detector may be a very thin metal strip, not more than 10^{-3} mm wide. The change in deflection of a particle arising from a change in μ_{eff} of the order of a nuclear magneton can therefore result in the particle being lost from the collected beam. We thus see that if the radio frequency is kept fixed and H_C is slowly varied, a sudden drop in collected current, i.e. a resonance, indicates that the radio frequency is coinciding with the Larmor precession frequency at that particular value of H_C. From the equation for ω_L derived in section 8.5.1, we can write

$$\mu_I = \frac{|I|\omega_L}{H_C}.$$

To appreciate the accuracy with which ω_L, and hence μ_I, can be determined we require to know the frequency range in which the resonance may be expected to take place. If we assume that μ_I is of the order of a nuclear magneton, that $H_C = 1000$ G and I is of the order of unity, then ω_L is found to be approximately 5 MHz. In this frequency range a frequency determination to an accuracy of one part in a million is easily achieved. The accuracy of the dipole-moment measurement will then be determined by the accuracy with which the magnetic field H_C is known. If the magnetic field can be guaranteed to be constant, even although it is not known in absolute value, then very precise comparisons of dipole moments can clearly be made.

An alternative method of procedure known as the '*flop-in*' *technique* is to adjust magnet B to bring to the detector only those molecules which have undergone a particular change in magnetic state on passing through magnet C. Under these conditions the collected current off resonance will be very small and will rise sharply when the resonance is reached.

The resonance technique was originally developed with a view to investigating magnetic moments using non-paramagnetic systems, i.e. systems in which the electrons make no contribution to the total magnetic moment. One then expects the interpretation to be very simple, involving only the reorientation of the nuclear moment in the field H_C. In Figure 47 is shown the result of such an experiment on the HD molecule. The sharp resonance shown is believed to arise from the reorientation of the proton spin in molecules which are in a zero rotational state. From the figure the resonant field is seen to be 945·8 G. The frequency being 4 MHz, it follows that

$$\mu_p = \frac{\hbar\omega_L}{2H_C} = 14 \times 10^{-31} \text{ J G}^{-1} = 2\cdot77 \text{ nuclear magnetons.}$$

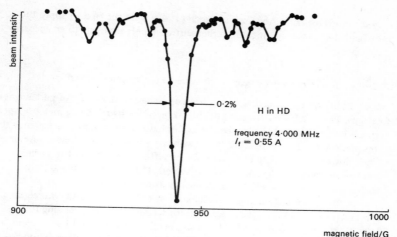

Figure 47 Resonance curve for HD molecules. The largest resonance corresponds to the transition of the proton spin for the zero rotational state

In pressing these measurements to the highest possible accuracy, account had to be taken of diamagnetic circulations induced in the electronic system. This may significantly modify the external magnetic field at the nucleus. Although corrections can be made from atomic theory, the uncertainty in these corrections is quite appreciable relative to the experimental errors.

The resonance technique has been successfully applied to the measurement of a large number of magnetic moments of nuclei whose spins had already been determined.

8.5.3   *The magnetic dipole moment of the neutron*

A technique similar in principle to the above has been applied to neutron beams. By scattering neutrons from a magnetized iron surface, or by transmitting them through thick sheets of magnetized iron, a degree of polarization can be produced in the beam. By this we mean that if n_+ is the number of neutrons with spins parallel to the magnetic field in the iron and n_- the number with spins antiparallel, then

$$P = \frac{n_+ - n_-}{n_+ + n_-}$$

may differ significantly from zero. A second mirror, or a second magnetized sheet, may be used to act as an analyser. The number of neutrons transmitted is then sensitive to any change taking place in the polarization of the beam as it travels between polarizer and analyser. Such a change in polarization can be produced by a long constant magnetic field containing an oscillator coil when the oscillator frequency equals the Larmor precession frequency of the neutron in the constant field. Alvarez and Bloch (1940) were the first to apply this difficult technique and measured the neutron dipole moment to an accuracy of 1 per cent.

8.5.4 Application of resonance beam technique to paramagnetic atoms

The resonance beam technique has been extended to the study of paramagnetic atoms. The much greater values of the magnetic moments being measured permit much lower field strengths to be used. This has the important consequence that magnetic states having practically the same μ_{eff} value in high fields, have in these lower intermediate fields μ_{eff} values sufficiently different to permit transitions between the states to be detected. This technique has greatly extended detailed knowledge of hyperfine structure.

By this method Rabi and his collaborators (1940) studied the moments of several nuclides. One of these was ^{7}Li, for which $J = \frac{1}{2}$, $I = \frac{3}{2}$ leading to $F = 2$ or 1 with M_F values as shown in Figure 48. In strong fields $\mathbf{J}$ and $\mathbf{I}$ are decoupled and the rules for transitions are $\Delta M_I = 0, \pm 1$; $\Delta M_J = 0, \pm 1$. In medium fields in which $\mathbf{J}$ and $\mathbf{I}$ remain coupled, the rules are $\Delta F = 0, \pm 1$; $\Delta M_F = 0, \pm 1$. Transitions involving a change in M_F are referred to as π-transitions; those in which M_F is not changed are referred to as σ-transitions. It is to be noted that in the present case of magnetic transitions the convention is the reverse of that used for electric dipole transitions. σ-Transitions occur only if there is an oscillating field component parallel to H_C and can be encouraged by the choice of coil geometry. This can be used to confirm the interpretation. From the observations of the resonant frequencies, the energy differences as a function of magnetic field can be obtained. Interpreting these on the basis of the theory discussed in section 8.1, and illustrated for the present values of J and I in Figure 46, one can make estimates of A and B from which the magnetic dipole and electric quadrupole moments can be calculated.

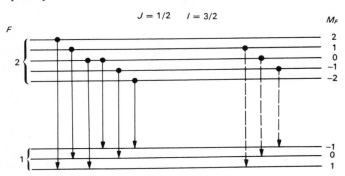

Figure 48 The magnetic levels, at intermediate field, of an h.f.s. doublet of an S$_{\frac{1}{2}}$ state. π-Transitions are shown in full lines, σ-transitions in broken lines

Magnetic resonant measurements on bulk materials

We now turn to the second group of resonant experiments which are concerned with the measurement of the moments of nuclei in solid or liquid materials. Whereas, in the beam techniques, the energy changes were being induced and observed in an isolated molecule, in the experiments we now have to discuss the molecules in the sample are in thermal contact with each other. Once more a magnetic field, which we shall here denote by H_0, is applied across the sample and a rotating field H is applied at right angles to H_0.

We first consider the case of a material containing only diamagnetic molecules having nuclei with non-zero spin. Assume the field H_0 to have been established but the field H not yet applied. The specimen will be affected only to the extent that the nuclei orient themselves at the allowed angles with respect to the direction of H_0. For simplicity we shall assume that the nuclear spin is $I = \frac{1}{2}$, although the arguments can be extended in an obvious way to other values of I. Let μ be the magnetic moment of the nucleus. There will then be a contribution of $-\mu H_0$ to the potential energy of the molecule when the nuclear spin is parallel to H_0 and $+\mu H_0$ when the spin and H_0 are antiparallel. When thermal equilibrium has been established, the populations of these two states are known from thermodynamic theory to be proportional to $\exp(\mu H_0/kT)$ and $\exp(-\mu H_0/kT)$ respectively, k being Boltzmann's constant and T the absolute temperature. Introducing n_+ and n_- to represent these populations we thus have

$$\frac{n_+}{n_-} = e^{2\mu H_0/kT}. \qquad\qquad 8.3$$

Let P_1 denote the probability that an individual dipole will in unit time, due to thermal disturbance, make a transition from the lower to the higher energy state (i.e. leave the n_+ population and join the n_-), and let P_2 denote the probability, also per dipole and per unit time, for a transition from the higher to the lower state. Since, when the system is in thermal equilibrium there must be as many transitions into as out of the lower energy state (and similarly for the higher), we must have

$$P_1 n_+ = P_2 n_-.$$

Hence $\quad \dfrac{n_+}{n_-} = e^{2\mu H_0/kT} = \dfrac{P_2}{P_1}.$

Now suppose the field H to be switched on. The probabilities P_1 and P_2 will both be increased by a constant amount, say P_3, which will represent the probability per unit time that a dipole change its orientation due to interaction with the radio-frequency field. Thus the populations of the states will be altered. If they now be denoted by n'_+ and n'_-, we can write

$$\frac{n'_+}{n'_-} = \frac{P_2 + P_3}{P_1 + P_3} = e^{2\mu H_0/kT'},$$

where we have introduced a new equivalent temperature, T', called the *spin temperature*, differing from T, the thermal temperature, since

$$\frac{n'_+}{n'_-} \neq \frac{n_+}{n_-}.$$

Adding a constant to numerator and denominator of P_2/P_1, which is greater than unity, will bring it closer to unity. Hence $T' > T$ and

$$\frac{n'_+}{n'_-} < \frac{n_+}{n_-}.$$

Thus there will be an increase in the population of the upper state at the expense of a decrease in the population of the lower.

8.5.6 *Nuclear magnetic resonance*

The energy for a transition from lower to upper energy state may now come either from the radio-frequency source driving the coil producing H or from the interaction of the nuclear moment with its surroundings, i.e. from the thermal energy of the specimen, which in this simplified treatment we regard as having infinite thermal capacity. Similarly the energy emitted when a transition from upper to lower state takes place may be delivered to the coil or to the thermal sink. The energy taken from the coil will be proportional to $P_3 n'_+$, while the energy delivered to the coil will be proportional to $P_3 n'_-$. Since n_- is always necessarily less than n'_+, there will be a net absorption of energy from the radio frequency supply. We note that

$$P_1 n'_+ < P_1 n_+ = P_2 n_- < P_2 n_-,$$

as a consequence of the change in population brought about by the application of the radio-frequency field. Thus more thermal energy is released by downward transitions than is absorbed by upward transitions. This surplus energy, which passes to the thermal sink, is of course exactly equal to the energy supplied by the radio-frequency source.

The difference in population between the upper and lower states, which is critical in the above considerations, is a very small fraction of the total population N. This can be seen to be so because

$$\frac{n_+ - n_-}{N} = \frac{n_+ - n_-}{n_+ + n_-} = \frac{n_+/n_- - 1}{n_+/n_- + 1} = \frac{2\mu H_0/kT - 1}{e^{2\mu H_0/kT} + 1}.$$

Now, for a magnetic moment of one nuclear magneton in a field of 10^4 G at 288 K,

$$\frac{2\mu H_0}{kT} = 2 \cdot 6 \times 10^{-6}.$$

Expanding the exponential function and taking only the first two terms we thus have

$$\frac{n_+ - n_-}{N} \simeq \frac{\mu H_0}{kT} = 1\cdot 3 \times 10^{-6}.$$

Despite this exceedingly small population difference the consequent net energy absorption by the process outlined is measurable. This effect was first demonstrated by Purcell and his colleagues (1946), using a radio-frequency cavity of volume 850 cm³ filled with paraffin. The cavity resonated at 29·8 MHz. When an external magnetic field was applied it was found that there was a resonant attenuation of the radio frequency as the magnetic field passed through 7100 G. The Larmor precession frequency of the protons in a field of that strength is thus established to be 29·8 MHz. A much simpler and more compact arrangement than that originally used has proved possible and is shown in Figure 49. A coil wound round a sample container of only a few cubic centimetres capacity is made part of a tuned circuit and incorporated in an apparatus which responds to a change in Q, the electrical selectivity of the circuit. The sample container is filled with water, or alternatively by another liquid under investigation. The radio frequency is kept constant. The field H_0 is then caused to vary slowly over a limited range by a subsidiary coil which carries a current alternating at 50 Hz. A cathode-ray oscilloscope trace, synchronized horizontally to the variation of H_0, can then be used to display the Q-variation vertically and a synchronized picture of the resonance is then achieved.

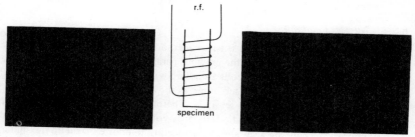

Figure 49 Schematic diagram of nuclear-magnetic-resonance measurement. Radio-frequency power is fed to a coil surrounding the specimen which is immersed in a uniform magnetic field

Apart from its importance in nuclear physics in providing an accurate measurement of the Larmor precession frequency for particular nuclei, and thus permitting an accurate determination of their magnetic moments, this technique has quickly become important in other fields of research. It enables precise comparisons to be made of steady magnetic fields. Providing the magnetic field is constant over the volume of the sample, so that all the protons in the sample have the same Larmor precession frequency, a very sharp resonance is obtained. From the central frequency of the resonance the field can be calculated in terms of the proton moment. As previously mentioned, frequencies in the range of ten megahertz, which is the relevant frequency for magnetic fields of a few thousand gauss, can readily be measured to one part in a million. Thus magnetic fields can be compared to this order of accuracy.

The technique has also provided considerable insight into the thermal properties of the samples under investigation. If the equilibrium condition is upset, for example by a sudden alteration of H_0, then the new equilibrium condition is approached with a time factor given by $1 - \exp(- t/T_1)$ and T_1 is called the *spin lattice relaxation time*. It is a measure of the tightness of coupling of the nuclear spins with the other degrees of freedom of the system. In a crystal structure T_1 may be of the order of seconds or even minutes. In a liquid it is more likely to be 10^{-2} s. It is greatly affected by the presence of a relatively small number of paramagnetic atoms. Advantage of this can be taken when water is the sample under investigation by adding a small amount of a paramagnetic salt, say manganese sulphate. This has the effect of reducing the relaxation time and sharpening the resonance.

The magnetic field measured by the nuclear-magnetic-resonance technique is of course the resultant of the external field and any internal fields in the material. The internal fields depend on the molecular structure, and may differ from the site of one proton to another in the same molecule. The pattern resulting from slightly shifted resonance can be a useful means of identifying the presence of a particular molecular species. As a consequence the technique, which is referred to as nuclear spin resonance (NSR), has already found a role in chemical investigations.

8.5.7 Nuclear induction

An alternative and equally sensitive method of detecting the resonance phenomenon discussed in the previous paragraph was developed by Bloch and co-workers in 1946. Two coils are set up with their axes at right angles to the constant field H_0 and to each other, as shown in Figure 50. One coil is fed with radio-frequency current to set up H as before. The other, the detector coil, is connected to a sensitive radio-frequency voltage detector. In the absence of a sample, the coils are accurately positioned to minimize magnetic-flux linkage, so that the signal induced in the detector coil is a minimum. In terms of the concept of two oppositely rotating fields one should rather say that there are two almost equal and opposite signals induced.

Now the sample to be investigated is introduced. We can consider H', the resultant of H_0 and that rotating field H which constitutes the synchronous perturbation, as defining the direction with respect to which the nuclear spins orient themselves. The nuclear dipole moments will have finite components parallel to H and sweeping round with H. These will induce a signal in the detector coil. If the two magnetic substates existing when $I = \frac{1}{2}$ were equally populated, then the net signal would be zero, as there would be as many positive as negative components contributing. In so far as the populations are not equal there will be a resultant signal. This signal is detected and amplified.

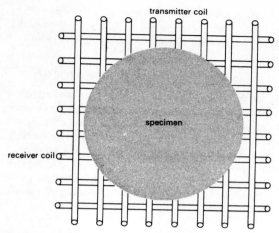

specimen

receiver coil

Figure 50 Schematic diagram of nuclear induction experiment. Radio-frequency power is fed to a transmitter coil. A radio-frequency signal is picked up by the receiver coil from nuclei precessing in the presence of a uniform magnetic field applied perpendicularly to the plane of the diagram

With this technique there is the additional information of the phase of the signal in the detector coil with respect to the phase of the voltage applied to the drive coil. This relative phase will depend on the sign of the magnetic moment as it would be altered by 180° if the direction of the Larmor precession were to be reversed.

8.5.8 *The measurement of electric quadrupole moments in bulk materials*

When a nucleus has a quadrupole moment and when it is situated in an internal electric field gradient $\partial^2\phi(0)/\partial z^2$ in the specimen, then there will be an electric quadrupole contribution to the total potential energy. This will give rise to a splitting of the resonances described above. In a crystal, where the electric field gradient is in a direction fixed with respect to the axis of the crystal, this splitting will vary with crystal orientation. Thermal motions giving rise to fluctuations in field gradient will broaden the lines and may make their resolution impossible. Also, where a measurement is possible, it is

$$B = eQ \frac{\partial^2\phi(0)}{\partial z^2}$$

which is measured and $\partial^2\phi(0)/\partial z^2$ is not usually known with any certainty. As pointed out above, the need to know the field gradient can be avoided by limiting the discussion to the ratios of quadrupole moments of two isotopes. By this method Becker (1951) measured the ratio of the quadrupole moments of ^{63}Cu and ^{65}Cu using single crystals of $K_3[Cu(CN)_4]$.

Pure quadrupole resonances have also been observed when a perturbing magnetic field applied perpendicular to the axis of the electric field in the absence of a field H_0 has induced transitions leading to absorption of radio-frequency power.

8.5.9 *The measurements of quadrupole moments by paramagnetic resonances*

The technique of Purcell described in section 8.5.6 has been applied to paramagnetic atoms. The much larger magnetic moment of the atom places the Larmor frequencies, and hence the resonant frequencies, in the microwave range. The effect of the nuclear spin is then to produce hyperfine splitting in the atomic resonances. By observing the degree of splitting as a function of H_0, diagrams of the type encountered in optical hyperfine splitting (see Figure 43, p. 137) are constructed. By comparing these with the theory of hyperfine splitting outlined above, the effect of the quadrupole moment can be detected and values of B estimated.

8.6 Summary

Spins of nuclei can be measured by several techniques and can be regarded as very well established for ground states of stable nuclei. In some cases the techniques can be applied to give a direct measurement of spin for unstable nuclei of suitably long half-life.

Magnetic moments can be measured with high precision, a precision limited in practice by the accuracy with which the internal and induced field produced by the orbital electrons at the nucleus can be calculated.

With respect to quadrupole moments the situation is very much less satisfactory. In certain cases B can be measured directly; in other cases it can be measured as a deviation of hyperfine structure from that expected on the basis of magnetic dipole moments alone. However, the deduction of Q from B is very uncertain because of the lack of knowledge of the internal electric field gradients at the nucleus. The ratio of Q for two isotopes is known with better accuracy on the assumption that the internal electric field gradient is the same for both isotopes.

As we shall see later, information about both electric and magnetic moments of nuclei comes from the study of transitions between their excited states. It is to that source that at present one has to look for better information about the electric quadrupole moment.

Values of spins, magnetic dipole moments and quadrupole moments for stable nuclei are listed in Appendix A.

Chapter 9
The Collective Model

9.1 Introduction

We recall that the discussion of section 6.10 showed that a good account of the ground-state spin of nuclei can be given in terms of the shell model. In that account it is assumed that nucleons of the same kind form pairs, their angular momenta coupling so that the resultant angular momentum of the pair is zero. In the case of (even, even) nuclei there is complete pairing and hence the spin predicted by the model is zero. This, without known exception, is in agreement with measured spins. In the case of odd-A nuclei there is always a nucleon of one type left unpaired. The nuclear spin then is assumed to arise entirely from the motion of this unpaired nucleon. In the case of (odd, odd) nuclei there is an unpaired nucleon of each kind and the nuclear spin has a contribution from the motion of each of these. The satisfactory agreement of this account of nuclear spins with experimental observations argues strongly for the validity of the shell model.

The attempt to extend the ideas of the shell model to explain magnetic dipole moments met, as we saw in section 7.3, with only limited success. It is true that (even, even) nuclei have zero magnetic moment, as well as zero spin, as would be expected on the assumption that the nucleons form pairs. However, in the case of odd-A nuclei, we see from Figures 31 and 32 (pp. 110–12) that, with few exceptions, the measured magnetic dipole moments are significantly smaller than the 'single-particle' predictions. One must conclude that in these cases the paired nucleons in the 'core' are not exactly compensating each others' magnetic moments but are making a contribution to the total dipole moment, this despite the fact that they do compensate each others' angular momentum. It is however to be noted that the main contribution to the magnetic moment still arises from the single particle.

The situation *vis-à-vis* the predictions of the shell model and the experimental facts is less satisfactory when we turn to electric quadrupole moments. The quadrupole moment is taken as a measure of the departure of the charge distribution from spherical symmetry. If a nucleus has a closed shell of protons it has no total angular momentum and hence no distinctive axis. It is therefore expected to exhibit spherical symmetry of charge. If we now take the case of one proton outside a closed shell then, unless that proton be in an s-state, its equivalent charge distribution will not be spherically symmetric and a nuclear electric quadrupole moment would be expected to result. Consider now a second

proton outside the shell. If this pairs with the first to give zero total angular
momentum, then again, on the above argument, spherical symmetry is restored.
If a third proton be added then, providing it does not split the existing pair, its
situation will be as for the first proton, that is, it should give rise to a quadrupole
moment calculable from its wave function. This argument can be extended to show
that spherical symmetry is to be expected when the number of protons is even and

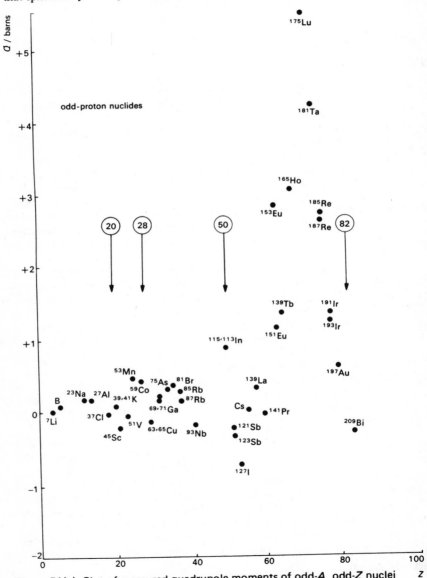

Figure 51(a) Plot of measured quadrupole moments of odd-A, odd-Z nuclei

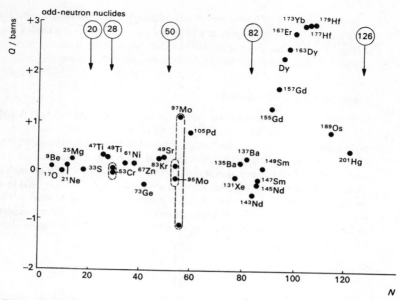

Figure 51(b) Plot of measured quadrupole moments of odd-*A*, odd-*N* nuclei

a departure from symmetry of charge distribution, associated with one proton, is to be expected when the proton number is odd. Now in a nucleus of radius, say, eight fermis, a single proton can at the most give rise to a quadrupole moment of 0·128 barns (this follows from the application of equation 7.12 to this simple situation where $r = z$). On the other hand, quadrupole moments as large as eight barns have been measured. Although these measurements, as was pointed out in section 8.6, are subject to considerable experimental uncertainties, nevertheless it is clear that there is an order-of-magnitude discrepancy at least between the single-particle-model prediction and the measurements. It is also to be noted from Figure 51 that the discrepancy is greatest when the proton number is midway between closed shells. As a further departure from shell-model expectations, odd-*A* nuclei having paired protons but an unpaired neutron exhibit quadrupole moments.

Nuclei which have shells half-filled not only show serious departure from shell-model predictions in respect of quadrupole moments, but have as a characteristic an excited state, not much above the ground state in energy, de-exciting by electric quadrupole transition to the ground state. The transition rate is higher than would be expected from single-particle considerations and is an indication that a large electric quadrupole moment is involved in the process. These considerations led Bohr and Mottelson (1953) to develop the collective model of the nucleus, which we now proceed to outline. This model attempts to extend the shell model rather than to replace it. It adds to the shell model some features of the liquid-drop model and, as it borrowed from both of these extreme nuclear models, it was originally sometimes referred to as the *unified model*.

In setting up the collective model a close analogy was maintained with the behaviour of the diatomic molecule as described by quantum mechanics. To make it clear how the ideas and nomenclature arose, we begin by outlining briefly the relevant features and properties of the diatomic molecular system.

9.2   Theory of the diatomic molecule

In Figure 52 we represent a simple diatomic molecule having nuclei at A and B. The theoretical treatment assumes that the energy of the system may be represented as the sum of three terms. The first of these is the energy of rotation of the system about the axis PQ; the second is the energy of vibration of the nuclei which, if their separation d is disturbed, are assumed to vibrate with simple harmonic motion about their equilibrium separation; the third and final term is due to the energy of the orbital electrons associated with one or other of the nuclei or orbiting both. The moment of inertia about the axis AB is assumed to be negligibly small and consequently no allowance is made for rotation about AB.

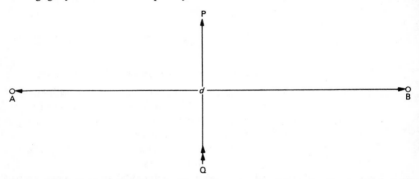

Figure 52   Schematic diagram of a diatomic molecule with atoms situated at A and B

We first consider the rotational energy about PQ. If the moment of inertia about the axis is $\mathscr{I}$ then we can write

$$E_{rot} = \tfrac{1}{2}\mathscr{I}\omega^2,$$

where ω is the angular velocity of the system about the axis PQ. The angular momentum about that axis will be $\mathscr{I}\omega$ and we note that, from classical mechanics, we can write

$$E_{rot} = \frac{(\mathscr{I}\omega)^2}{2\mathscr{I}}.$$

When the system is treated as a quantum-mechanical rotator, we assign to it, when it is in a stationary state, a quantum number R which is related to the quantized angular momentum of rotation by the equation

Angular momentum about axis of rotation $= \sqrt{[R(R + 1)]}\,\hbar$.

In the quantum-mechanical treatment we then have

$$E_{rot} = \frac{R(R+1)\hbar^2}{2\mathscr{I}}.$$

9.1

If now the molecule has an electric-dipole moment it will be capable of interacting with the electromagnetic field. It may therefore radiate energy or it may absorb energy from the field thereby making a transition between states differing by ± 1 in R-value. The energy of quantum radiated or absorbed will be given by

$$(h\nu)_{rot} = \Delta E_{rot} = \frac{[R(R+1)-(R-1)R]\hbar^2}{2\mathscr{I}} = \frac{R\hbar^2}{\mathscr{I}}.$$

9.2

Let us take the HCl molecule as an example to which this theory should apply. The separation d will be of the order of 10^{-10} m. Hence, taking the moment of inertia about an axis through G, the centre of mass, perpendicular to the line joining the nuclei we have

$$\mathscr{I} = (M_C x_C^2 + M_H x_H^2) = \tfrac{35}{36} M_H d^2 \simeq 1 \cdot 6 \times 10^{-44} \text{ g m}^2.$$

Hence $\nu_{rot} = \dfrac{R\hbar}{2\pi\mathscr{I}} \simeq R \times 10^{+12}$ Hz.

This frequency, for small values of R, corresponds to the far infrared region of the electromagnetic spectrum. When this spectral region is examined in the absorption spectrum of HCl, a sequence of lines, listed by wave number (i.e. $k = 1/\lambda = \nu/c$) in Table 5, is found. The difference in wave numbers, apart

Table 5: Absorption Spectrum of HCl in the Far Infrared.

k (observed) x mm	Δk x mm	$\dfrac{k}{\Delta k}$	k (calculated) x mm	[k (calculated) $-$ k (observed)] x mm
8·303	2 x 2·063	4·06	(8·303)	—
...				
12·430	2·073	6·07	12·455	0·025
14·503	2·048	7·08	14·530	0·027
16·551	2·035	8·08	16·606	0·055
18·586	2·052	9·08	18·682	0·096
20·638	2·012	10·08	20·758	0·120
22·650		(11·06)	22·833	0·183

from that between the first and second entries, is seen to be effectively constant. The exception is so close to twice this constant difference as to suggest that there is a line missing. If the interpretation in terms of rotation is correct, then

$$k_{rot} = \frac{\nu_{rot}}{c} = \frac{R_{rot}\hbar}{2\pi\mathscr{I}c}$$

and hence $\Delta k_{rot} = \dfrac{\hbar}{2\pi \mathscr{I} c}$.

It follows that

$$\frac{k_{rot}}{\Delta k_{rot}} = R_{rot},$$

where R_{rot} should be integral.

From column three of the table it can be seen how well this prediction is borne out. If we now take the nearest integers and use the first wave number we can calculate the wave numbers of the other lines. It can be seen that the agreement is good but that there is a discrepancy growing systematically as shown in the final column of the table. This discrepancy is understandable in terms of the simple rotator model. As we proceed to higher rotational states, the centrifugal force will increase and this will tend to stretch the molecule. The increased separation will lead to an increase in $\mathscr{I}$ and, according to equation 9.1, a decrease in the expected value of E_{rot}.

We can use the spectral information to find accurate, rather than order-of-magnitude, values of I and d. We have

$$\mathscr{I} = \frac{R\hbar}{2\pi \nu_{rot}} = \frac{R\hbar}{2\pi c k_{rot}} = 2\cdot 7 \times 10^{-44} \text{ g m}^2.$$

It follows that $\quad d = \sqrt{\dfrac{36\mathscr{I}}{35 M_H}} = 1\cdot 29 \times 10^{-10} \text{ m}.$

There is thus seen to be impressive agreement between the spectroscopic data and the predictions of the rotator model.

The existence of rotational states should affect the thermodynamic properties of a gas consisting of diatomic molecules. From the equipartitioning of energy in kinetic theory, it follows that C_v, the specific heat of a gas, is $\frac{1}{2}nR$, where R is the gas constant and n the number of degrees of freedom possessed by the molecules. If the rotational states enter into the equipartitioning of the energy they will add two additional degrees of freedom, since rotation can take place independently about two axes perpendicular to AB in Figure 52. The extent to which the rotational states take up thermal energy will be dictated by the Boltzman factor. This ensures that the rotational states are negligibly populated when

$$kT \ll (h\nu)_{rot} \simeq \frac{\hbar^2}{\mathscr{I}}.$$

When however $\quad kT \simeq \dfrac{\hbar^2}{\mathscr{I}} \quad \left(\text{i.e. } T \simeq \dfrac{\hbar^2}{\mathscr{I}k} = 30 \text{ K} \right),$

then n should increase from 3 to 5. In fact C_v is observed to increase from $\frac{3}{2}R$ to $\frac{5}{2}R$ in this temperature range, further supporting the validity of the rotator model.

We now turn to the vibrational energy. If the 'restoring' force is proportional to the change in the separation distance of the nuclei, then the system should correspond to a quantum-mechanical oscillator. Such an oscillator has a set of stationary states of energy given by

$$E_{vib} = (n + \tfrac{1}{2})(h\nu)_{vib},$$

9.3

where n is an integer. A knowledge of the restoring-force constant would enable $(h\nu)_{vib}$ to be calculated. However, rather than trying to relate the restoring force to the separation distance through atomic theory, we turn to thermodynamics to see whether there is evidence as the temperature increases for the appearance of further degrees of freedom which might be connected with the population of vibrational states. As shown in Figure 53 there is a rise from $\tfrac{5}{2}R$ to $\tfrac{7}{2}R$ in C_V in

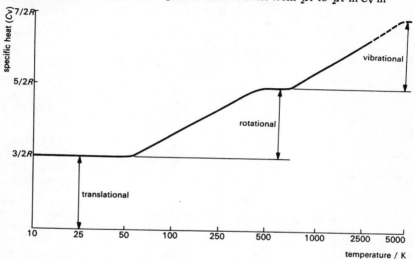

Figure 53 Representation of the variation of specific heat of molecular hydrogen with absolute temperature

the range of temperature around 1500 K. We interpret this rise as being due to the two additional degrees of freedom, one, according to kinetic theory, for the potential energy and one for the kinetic energy, entering as the vibrational states become populated. From the temperature at which the increase takes place we can estimate the vibrational frequency and find

$$\nu_{vib} = \frac{kT}{h} = 3 \cdot 12 \times 10^{18} \ s^{-1}.$$

Hence the wave number will be given by

$$k_{vib} = 1 \cdot 04 \times 10^2 \ mm^{-1}.$$

We deduce therefore that transitions between vibrational states may give rise to the emission or absorption of quanta in the near infrared. We note that according to equation 9.3 vibrational states are expected to be equally spaced.

Finally, we have the electronic contribution to the molecular energy. As far as the electrons are concerned the situation is not appreciably different from that existing in a simple atom. The electronic states are therefore expected to be such that transitions between them will give rise to quanta of radiation in the visible region of the electromagnetic spectrum.

The description of the diatomic molecule in terms of rotational, vibrational and electronic states fits experimental observations very well. Transitions may be observed directly between rotational levels or directly between vibrational levels; alternatively the existence of these levels may be deduced from the structure that is observed when transitions between electronic states are measured with high resolution. The relationship between the states is illustrated in Figure 54.

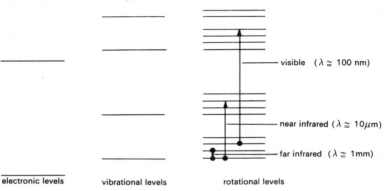

electronic levels   vibrational levels   rotational levels

Figure 54 Relationship of electronic, vibrational and rotational levels in diatomic molecule

It should be noted that one reason for the description working so successfully is that the characteristic frequencies associated with the three energy contributions are well separated.

9.3   An outline of the collective model of the nucleus

We start by considering a nucleus in which the nucleons are arranged in closed shells. Such a nucleus has spherical symmetry and, should a deformation be impressed on it, forces will act to restore the symmetry. If now a few nucleons be added to form a heavier nucleus, then the shell model is looked on as still providing an adequate description of the nuclear behaviour. Nucleons form pairs which also have spherical symmetry. However, as the process proceeds and the shell is progressively filled, the number of pairs of nucleons increases and the nucleons in the partly filled shell form a 'softer' structure, the restoring forces coming into play when the surface is deformed, being weaker than is the case for a closed shell. Such nuclei are referred to as *transitional nuclei*. As more nucleons are added to the shell, the 'pairing' effect is assumed to become less able to maintain symmetry and the partly filled shell passes through the state of being an easily deformed sphere and becomes a spheroid or ellipsoid of revolution. Nuclei of this type are called *deformed nuclei*. The subtraction of nucleons from a closed shell can be treated as readily by the shell model as the addition of nucleons, assignments of spins and parities to the 'vacancies' proceeding as for nucleons. Thus as we go on adding nucleons to deformed nuclei we pass into a further range of transitional nuclei before reaching the next closure of the shell. We proceed to develop the consequences of these 'properties' we have conferred on the nucleus.

9.4 Rotational states of deformed nuclei

We begin by considering an (even, even) deformed nucleus, i.e. one having an approximately half-filled shell of nucleons. The shape is, as discussed above, no longer that of the spherical closed-shell nucleus but has become that of an ellipsoid of revolution. The extent of the deformation is measured by the ellipticity

$$\epsilon = \frac{c - b}{R_0},$$

where c is the length of the semi-axis of symmetry and b the length of the other equal semi-axes. R_0 is $\frac{1}{2}(c + b)$. If we take the result derived in section 7.7 for the quadrupole moment of a deformed sphere and use the values of ϵ and R_0 defined above, we can write for the intrinsic quadrupole moment of the nucleus we are now discussing

$$Q_0 = Z\tfrac{2}{5}(c^2 - b^2) = \tfrac{4}{5}Z\epsilon R_0^2.$$

In the present case the nucleus in its ground state has zero angular momentum. Hence, if a direction is specified by an electric field gradient, the spheroid will orient itself randomly, leading, as discussed in section 7.9, to an effective quadrupole moment of zero.

In analogy with a diatomic molecule we now consider the possibility of the nucleus having states in which it rotates as a quantum-mechanical rotator about the axis PQ in Figure 55. This rotation takes place with low enough angular velocity to permit the single-particle orbits, which involve much higher angular velocities, to follow the rotation of the spheroid shape. If, as for the diatomic molecule, we associate a quantum number R with the rotation, the angular momentum about the axis PQ will be given by $\sqrt{[R(R + 1)]}\,\hbar$ and the energy will be given by

$$\frac{R(R + 1)\hbar^2}{2\mathscr{I}},$$

where $\mathscr{I}$ is the moment of inertia of the nucleus about the axis PQ.

We now define two sets of orthogonal axes, one set, x, y, z fixed with respect to the nucleus, the z-axis being the axis of symmetry, and a set x', y', z' fixed in the laboratory. If we take z' as a specified direction then the nucleus will orient

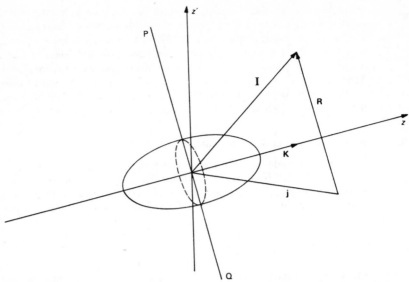

Figure 55 Spheroidal nucleus with intrinsic rotation about z-axis and capacity for relatively slow rotation about the axis PQ

itself to have a set of magnetic substates, the maximum angular momentum about the z'-axis being $R\hbar$. In this case R is to be equated to I, the spin of the excited state of the nucleus. The rotational states are therefore expected in analogy with equation **9.1** to have an energy spacing given by

$$\frac{I(I+1)\hbar^2}{2\mathscr{I}}.$$

Since in the case being discussed the spin of the ground state is zero, we note that the rotator is such that a rotation of $180°$ about the axis x or the axis y leaves the system unaltered. In this situation, as for a diatomic molecule with identical nuclei, the states of odd spin are missing and all the states have even parity. The expected states then have spins and parities given by $0^+, 2^+, 4^+, 6^+, \ldots$, with the energy spacings proportional to $I(I+1)$.

It has long been known that, with few exceptions, and these usually associated with closed-shell nuclei, the first excited state of (even, even) nuclei is a 2^+ state. The spacing of this state above the ground state diminishes steadily as the shell fills, reaching a minimum with the deformed nuclei. In Table 6 are entered sequences of low-lying states found in four (even, even) deformed nuclei, having spins and parities as predicted above. It is seen that there is a remarkable similarity in energy spacing as between the nuclei, and when the spacing is compared with the theoretical prediction we see that the agreement is good but that there is a growing discrepancy as I increases. We recall a similar behaviour in the case of

Table 6

Spin and parity	$^{170}_{72}\text{Hf}$	$^{164}_{66}\text{Dy}$	$^{164}_{68}\text{Er}$	$^{166}_{70}\text{Yb}$	$\dfrac{I(I+1)}{6}$
16^+	3·147(31·47)				45·33
14^+	2·564(25·64)				35·00
12^+	2·013(20·13)				28·17
10^+	1·503(15·03)		(1·466)(16·02)	2·172(21·33)	28·17
8^+	1·041(10·41)	0·839 (11·43)	1·024(11·19)	1·604(15·76)	18·33
6^+	0·641 (6·41)	0·501 32 (6·83)	0·614 (6·71)	1·097(10·77)	12·00
4^+	0·321 (3·21)	0·242 23 (3·30)	0·299 (3·27)	0·667 (6·55)	7·00
2^+	0·100 (1·00)	0·073 392(1·00)	0·0915(1·00)	0·330 (3·24)	3·33
0^+	0 (0)	0 (0)	0 (0)	0·1018(1·00)	1·00
				0 (0)	0

rotational spectra of molecules, discussed in section 9.2. We assume that the explanation is similar in this case, namely that as the rotational angular velocity increases, the increasing centrifugal force is stretching the nucleus thereby increasing the effective $\mathscr{I}$ and leading to a reduction in the energy spacing.

We now turn to the case of odd-A nuclei, which, on the single-particle view, have non-zero ground-state spin. Let us assume that the angular-momentum quantum number of the unpaired nucleon is j. This nucleon will give rise to a component of angular momentum along the z-axis. This component we denote by $K\hbar$. Consider now the rotational states based on this ground state. The angular momentum of rotation will add vectorially to the angular momentum $\sqrt{[j(j + 1)]}\hbar$ of the unpaired nucleon to give a resultant angular momentum $\mathbf{I}$ with $|\mathbf{I}| = \sqrt{[I(I + 1)]}\hbar$. The angular momentum of rotation is, from Figure 55, seen to be the component of $\mathbf{I}$ along the axis of rotation and to have a magnitude given by $\sqrt{[I(I + 1) - K^2]}\,\hbar$. It follows that the rotational energy in this case is given by

$$E_{\text{rot}} = \frac{[I(I+1) - K^2]\hbar^2}{2\mathscr{I}}.$$

Apart from the term $-K^2\hbar^2/2\mathscr{I}$, which will be the same for all the rotational levels in the sequence and therefore will not affect the spacing, we see that this is the same result as was found for (even, even) nuclei. However, the symmetry which suppressed the states with odd spins in that case is not present in the case of odd-A nuclei because K is no longer zero. Hence we expect the spins of the rotational states to be $K, K + 1, K + 2, K + 3, \ldots$, all half-integral. The parity of all the states is expected to be the same as the parity of the unpaired particle.

Nuclei having $K = \frac{1}{2}$ constitute a special case. The angular momentum about the z-axis is not necessarily unaffected by the rotational motion, and a more complicated treatment beyond the scope of the present discussion is necessary. We simply quote here the result of the theoretical analysis for the case of $K = \frac{1}{2}$, namely

$$E_{\text{rot}} = \frac{1}{2\mathscr{I}}\,\hbar^2[I(I+1) - K^2 + a(-1)^{I+\frac{1}{2}}(I + \tfrac{1}{2})], \qquad 9.4$$

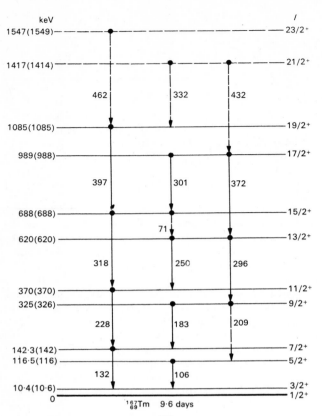

Figure 56 Rotational band of ^{167}Tm. The energies in parenthesis were calculated from a theoretical formula having four adjustable parameters

where the parameter a, called the *decoupling parameter*, depends on the details of the intrinsic nuclear structure. In Figure 56 are drawn a sequence of energy levels of ^{167}Tm. The theoretical estimates based on equation 9.4, with terms added to allow for the stretching in the higher rotational states, show how accurately the model can be made to match the experimental results.

The absolute spacing between states, together with the appropriate formula for E_{rot}, can now be used to calculate $\mathscr{I}$. The value thus found is considerably less than that which would arise were the nucleons in the nucleus to move around the z-axis with a common angular velocity. This would amount to a rigid-body rotation of the nucleus and would result in

$$\mathscr{I}_{rig} = \tfrac{2}{5}AMR_0^2,$$

where M is the nucleon mass.

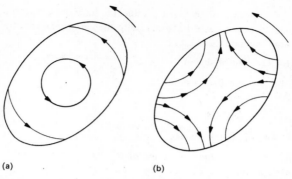

(a)   (b)

Figure 57 Rotation of spheroid (a) in which all parts move circularly round the
axis of rotation as for a rigid body and (b) in which the shape rotates but the
component parts oscillate along paths which do not circle the axis of rotation.
The latter wavelike rotation corresponds to *irrotational flow*

The measured values however are greater than would arise from a movement of
the nucleons depicted in Figure 57, corresponding to irrotational flow in which
the surface shape rotates although the individual nucleons do not have a simple
circular motion round the axis. In the case of irrotational flow the moment of
inertia would be given by

$$\mathscr{I}_{irrot} = \tfrac{2}{5}AMR_0^2\,\epsilon^2, \qquad \text{where} \quad \epsilon = \frac{c-b}{R_0}.$$

The actual behaviour of nucleons, to judge from the measured values of $\mathscr{I}$, lies
somewhere between the two extremes of rigid-body rotation and irrotational flow.

9.5   Vibrational states of nuclei

9.5.1   Corresponding to the linear vibration of the diatomic molecule, we have in the
case of the nucleus, the vibration of the three-dimensional sphere (in the case of
closed-shell nuclei) or spheroid (in the case of deformed nuclei). In the simplest
approach, we assume that the nucleus behaves as if it were an incompressible fluid
oscillating under the restoring force of surface tension when deformed in shape.
Just as for a vibrating elastic cord fixed at both ends, which is the two-dimensional
analogue and more easily visualized, the spherical system is capable of vibrating
with one of a series of harmonics or with, in the general case, a mixture of such
harmonics. In the case of the sphere, the harmonics are described by functions
$Y(\theta, \phi)$ expressing the displacement of a point (R, θ, ϕ) on the spherical surface
from its initial undisturbed position. The form of the functions $Y(\theta, \phi)$ has to be
found by solving the differential equation corresponding to elastic waves on the
spherical surface. This is mathematically similar to the solution of the Schrödinger
equation of section 6.3. In place of l and m introduced in the separation of the
variables in that analysis, we introduce in the present application λ and μ. As in
the case of a vibrating string, we expect the lowest-order harmonics (i.e. those

described by the lowest λ- and μ-values) to have the lowest frequencies and to be associated, in the nuclear context, with the *vibrational states* of lowest energy. We now examine the values of

$$Y_{\lambda\mu}(\theta, \phi) = \Theta_{\lambda\mu}(\theta) \Phi_{\mu}(\phi)$$

for the lowest-order harmonics.

We recall that

$$\Theta_{\lambda\mu}(\theta) = P_{\lambda}^{\mu}(\cos \theta)$$

and take $\Phi_{\mu}(\phi) = \cos \mu\phi,$

rather than the complex function of section 6.4. λ takes the value zero or a positive integer, while μ, for a given value of λ, takes positive or negative integral values (or zero) and $|\mu| \leqslant \lambda$. As we are mainly interested in visualizing the modes of vibration, our concern is with the angular dependence of the deformation and we therefore take as unity the constant normalization factor, which strictly speaking ought to occur in the spherical harmonic. The value of the normalization factor controls the relative strength of the particular harmonic being discussed.

(a) For $\lambda = \mu = 0$, $Y_{00}(\theta, \phi) = 1$.

This mode constitutes an isotropic expansion or contraction of the spherical shape and, as we are considering the fluid incompressible, the strength of this harmonic must be zero.

(b) For $\lambda = 1, \mu = 0$, $Y_{10}(\theta, \phi) = \cos \theta$.

The surface in this mode will be described by the equation

$$r = R_0(1 + A \cos \theta).$$

For small values of A this represents a movement of the whole sphere along the z-axis without any change in the shape of the surface.

(c) For $\lambda = 1, \mu = \pm 1$, $Y_{1 \pm 1}(\theta, \phi) = \sin \theta \cos \phi$.

For this mode the surface is described by the equation

$$r = R_0(1 + B \sin \theta \cos \phi).$$

Again this represents a bodily movement of the sphere without shape distortion. In this case the centre is displaced in a direction perpendicular to the z-axis.

(d) For $\lambda = 2, \mu = \pm 1$, $Y_{2 \pm 1}(\theta, \phi) = 3 \sin \theta \cos \theta \cos \phi$.

This vibration, while symmetric with respect to reflection in the origin ($\phi \rightarrow \phi$, $\theta \rightarrow \pi + \theta$), is not symmetrical with respect to reflection in the plane of the x- and y-axes ($\phi \rightarrow \phi, \theta \rightarrow \pi - \theta$); on the same basis as the rotational states with $I = 1, 3$, etc., were discounted above, this mode of vibration is also now discounted.

(e) For $\lambda = 2, \mu = \pm 2$, $Y_{2 \pm 2}(\theta, \phi) = 3 \sin^2 \theta \cos 2\phi$.

This vibration has the necessary symmetry. If we think of the 'poles' of the nucleus lying on the z-axis with the xy plane as the equatorial plane, then, as shown in Figure 58, this mode of vibration will involve transport of nuclear mass around the

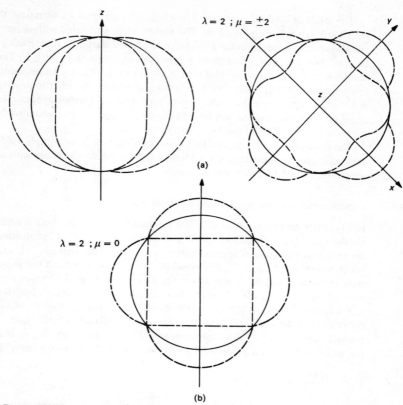

Figure 58 (a) γ-Vibration in which a wave runs equatorially with no variation of polar diameter. (b) β-Vibration in which the polar diameter oscillates with no variation in shape of section at different values of ϕ

equator. This will involve angular momentum about the z-axis amounting in the case of this second-order harmonic to $2\hbar$ units.

(f) For $\lambda = 2$, $\mu = 0$, $Y_{20}(\theta, \phi) = \frac{1}{2}(3\cos^2\theta - 1)$.

This vibration also is seen to have the necessary symmetry. From Figure 58 it may be seen that in this mode the transport of mass is towards and away from the poles, i.e. in a plane of constant ϕ. This does not involve the generation of angular momentum about the z-axis.

The behaviour of the system with respect to angular momentum is therefore seen to be the same as for a particle of spin 2, with λ playing the role of the spin and μ playing the role of the magnetic quantum number. The permitted values of the magnetic quantum number in this case being 2 and 0. The energy of vibration is said to be carried by a *phonon* on which we confer these angular momentum properties.

The mode of vibration with $\lambda = 2$, $\mu = 0$, is referred to as a β-vibration; that with $\lambda = 2$, $\mu = \pm 2$ is a γ-vibration. The energy levels associated with these vibrations will be spaced as for a simple quantum oscillator, i.e. the spacing will be even. The first level, a one-phonon level, will have spin and parity 2^+. The second level will contain two phonons, each of spin 2. The angular momentum of one phonon will couple with the angular momentum of the other to produce a resultant angular momentum which must have symmetry with respect to the plane xy. This requirement will limit the resultant angular momentum to the values 0, 2, 4. In all cases the parity will be positive.

The discussion can be continued to take into account more complicated harmonics, for example taking $\lambda = 3$ will involve octupole vibrations with angular momentum equal to 3 and negative parity.

9.5.2 *Experimental evidence for vibrational states*

In Figure 59 the energy levels of ^{106}Pd are drawn. We see that there is evidence for the existence of three levels with spins and parities 0^+, 2^+, 4^+ at an energy above the ground state of approximately twice that of the first excited state, which has spin and parity 2^+. (We recall that for a rotational band the second state in the sequence has a spin and parity 4^+ and an energy 3·33 times that of the first excited state.) We also note that the upper 2^+ state decays preferentially to the lower 2^+ state rather than to the ground state. This is evidence of the operation of a selection rule typical of harmonic oscillators, namely that transitions between neighbouring states are favoured. We notice that in two respects the theoretical predictions are not accurately borne out, namely that

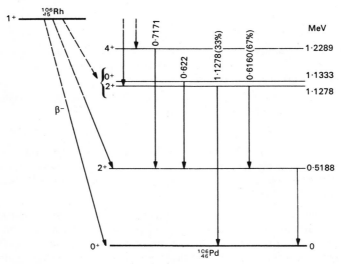

Figure 59 Vibrational states in ^{106}Pd

the two-phonon states are not degenerate and that the energy spacings are not precisely in the ratio 2:1. It has to be remembered that the theory is based on the assumption that the amplitudes of oscillation are within the range of linearity of the restoring forces; this assumption may not be strictly valid.

Many other examples of vibrational states have been found. A study of these reveals that as closed shells are approached the phonon energy increases, indicating that the rigidity of the spherical shape is increasing.

The situation in the case of the nucleus is more complex than it was for a diatomic molecule. The energies of the single-particle states are much closer to the vibrational-state energies than were the electronic-state energies in the case of the molecule. There is thus less validity in the picture of the fast-nucleon orbitals following the relatively slow vibrations of shape of the nucleus. This means that when we come to consider spherical, or nearly spherical, nuclei having odd A-values there is the complication that the unpaired nucleon has its motion coupled with the vibrations of the core. The results in such cases are then difficult to interpret.

Vibrational states can also arise in deformed nuclei; then the spheroid must be imagined as vibrating as well as rotating. In that case phonons will give rise to vibrational structure which will complicate the rotational bands.

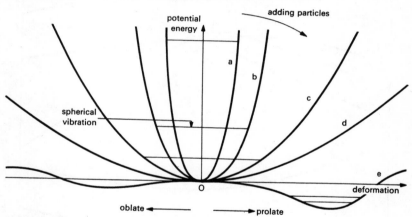

Figure 60 Schematic representation of potential energy as a function of deformation for nuclei ranging from closed shell (a) to permanently deformed (e)

In Figure 60 a schematic representation of the potential energy of the nucleus as a function of deformation summarizes the points that have been made with respect to vibrational levels.

9.6 Intrinsic states

In addition to rotational and vibrational states we still have to consider the states of the individual nucleons moving in the average nuclear potential. These correspond to the electronic states in the diatomic molecule. In the case of a spherical nucleus (i.e. where the shells are filled or almost filled), the individual nucleons experience, as in the simple shell model, a spherically symmetric potential. In the case of the deformed nuclei (i.e. where there is an approximately half-filled shell) the potential will not be spherically symmetric, and the results of section 6.4 will no longer be strictly valid. The analysis of that section has been repeated for an ellipsoidal potential with the results shown in Figure 61. We note that there is further removal of the degeneracies associated with the single-particle states.

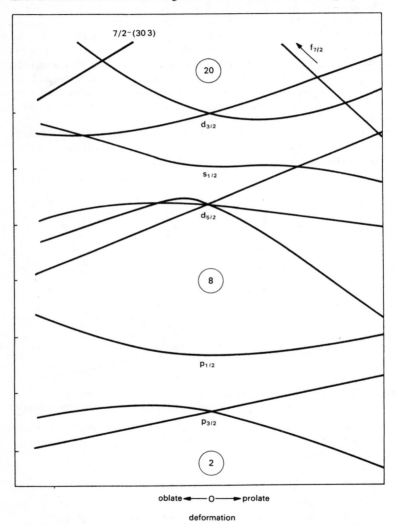

Figure 61 Variation of nuclear energy levels with deformation of nucleus (Nilsson diagram)

9.7 Summary

In marshalling the large amount of information now available about the excited states of nuclei we are assisted by the above considerations to the following extent. When we are dealing with a nucleus having closed shells, or almost closed shells, we can seek to interpret the pattern of levels in terms of vibrational states and single-particle states. The latter should be well described by the simple shell model. When we are dealing with nuclei with half-filled, or nearly half-filled, shells, then we have to take into account the deformation of the nucleus. This means that we have the additional possibility of rotational states, and also that the single-particle states will be modified by the fact that the nuclear potential is now axially symmetric but not spherically symmetric.

Chapter 10
Excited States of Nuclei

10.1   **Introduction**

In this chapter we discuss the properties of excited states of nuclei. We consider the electromagnetic transitions that occur between states and relate the rates at which these transitions occur to the properties of the states concerned. The methods of measurement and experimental results obtained are briefly reviewed.

10.2   **Production of excited states**

The existence of excited states of nuclei was established, as we saw in section 3.15, by an investigation of the fine structures in α-particle spectra. The grouping of nucleons which constitutes the daughter nucleus following α-emission is not always left in its lowest energy state, i.e. in its ground state. Should it be left in a higher energy state, i.e. an *excited state,* then electromagnetic transitions involving the emission of photons take place as the system de-excites.

A similar effect, as noted in section 4.12, is observed in the case of certain β-emitters. In Figure 62 the energy levels in ^{249}Bk and ^{93}Zr, constructed from measured α- and β-spectra, are shown to indicate the number of levels that can be involved and the resulting complexity of the spectrum of γ-rays emitted in association with the α- and β-particles.

The mass–energy conditions in α- and β ·decay limit the energy range of excitation of the nucleus which is explored by these methods. Even within that limited range, the lower-lying states are favoured in production by the energy dependence of the α- and β-transition probabilities. Further it is a consequence of the selection rules for α- and β-decay that only those levels having certain spins and parities will appear in the decay schemes.

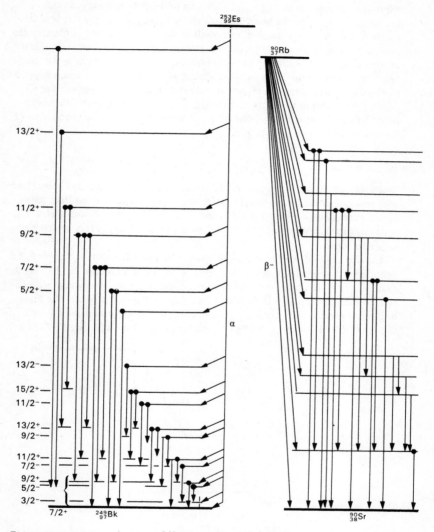

Figure 62 (a) Energy levels of ^{249}Bk deduced from α-particle spectrum emitted by ^{253}Es. (b) Energy levels of ^{90}Sr deduced from β^- spectrum of ^{90}Rb

10.2.1 Photoexcitation

Other methods are now available for the production of nuclear excited states, which enable the energy-level diagrams to be considerably extended. The exposure of a nucleus in its ground state to a beam of high-energy X-rays, produced by stopping the beam from an electron accelerator in a target of a material of high atomic number, can result in the absorption of a photon by the nucleus leading to a transition from the ground state to an excited state. This excited state can lie within an energy range limited only by the energy of the electron producing the X-ray bremsstrahlung spectrum. There are again selection rules involved in the transitions and it is therefore not to be expected that all excited states within the accessible energy range will be populated. We return later to discuss methods involving other nuclear reactions for the production of excited states.

10.2.2 Bound states

When the full energy range of excited states is being considered, the states divide into two categories. Those states whose excitation energies are less than the binding energy of the least tightly bound nucleon are termed *bound states*. Those states whose excitation energy exceeds the nucleon binding energy are termed *unbound states*.

Usually a bound state promptly de-excites by γ-ray emission, a process to be considered in some detail below. However, as discussed in section 6.11, there are examples of γ-ray transitions, the so-called isomeric transitions, which proceed at such a slow rate that β-transitions to a neighbouring isobar, where the mass difference permits this process, successfully compete. An interesting example of this is the 2·83 hour state 388 keV above the ground state of ^{87}Sr, which largely

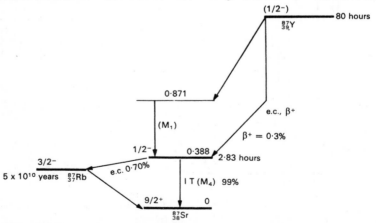

Figure 63 Decay scheme to show the competition between electron capture and γ-ray emission from the isomeric state in ^{87}Sr

de-excites by γ-ray emission to the stable ground state, but in 0·7 per cent of the decays proceeds by electron capture to the ground state of ^{87}Rb. The decay scheme is shown in Figure 63. This deviation from the normal γ-ray de-excitation is however very rare.

The unbound states of the nucleus have no counterparts in the atomic system. In the case of the atom, as the first ionization energy is passed a continuum of energy states is entered. There is no resonant absorption of photons from the neutral atom ground state into states in this continuum. In the case of the nucleus the behaviour is quite different. Resonant absorption is observed from the ground state to states which lie above the nucleon binding energy. The resulting states may decay by γ-emission but this process usually competes unsuccessfully with nucleon emission, as will be discussed below.

10.2.3   *Coulomb excitation*

Knowledge of excited states has been notably extended in recent years by the exploitation of *Coulomb excitation*. A heavy particle, either a proton or a heavy ion of suitable energy, passing a nucleus in its ground state at a distance greater than the effective range of nuclear forces, gives rise to a transient electric and magnetic field at the nucleus. The resulting perturbation may induce a transition of the nucleus to an excited state. This process, like γ-ray absorption, has the attraction of being entirely electromagnetic in nature and therefore calculable with the aid of the theoretical equipment, in the form of quantum electrodynamics, developed and tested in the field of atomic physics. Coulomb excitation has been particularly successful in demonstrating the existence and determining the properties of the low-lying states associated with rotational bands (see section 9.4). Another technique which has proved extremely valuable in recent years in the study of excited states is electron scattering. A beam of monoenergetic electrons is directed at a target nucleus. The electrons are scattered by purely electromagnetic processes, the scattering events falling into two categories, elastic and inelastic scattering. In an elastic scattering event the nucleus recoils to conserve linear momentum but does not make a transition from its ground state. A comparison of the observed angular distribution of the scattered electrons with that calculated from relativistic quantum electrodynamics, for a given nuclear charge distribution, provides valuable information on the nuclear charge radius. In the event of the nucleus making a transition from its ground state to an excited state the electron scattering is said to be *inelastic*. As in Coulomb excitation, but in this case using the fully relativistic treatment, the transition details are calculable from quantum electrodynamics.

10.2.4   *The compound nucleus*

Coulomb excitation and electron inelastic scattering are two specialized forms of nuclear reactions by means of which both bound and unbound excited states may be populated. In the more general form of nuclear reaction, where the bombarding

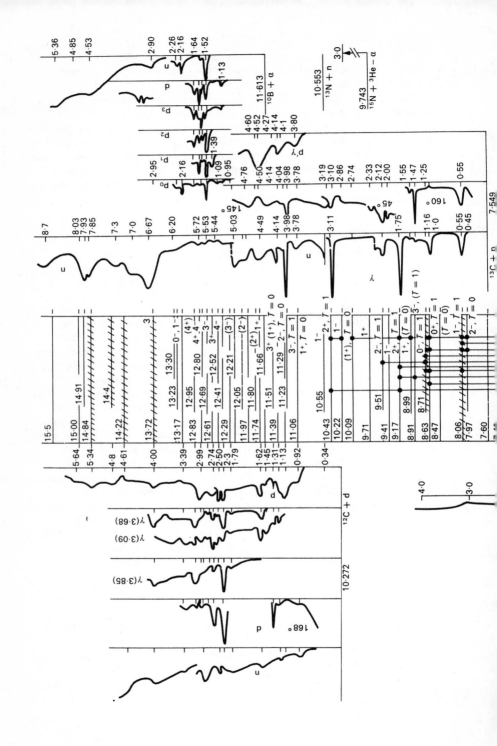

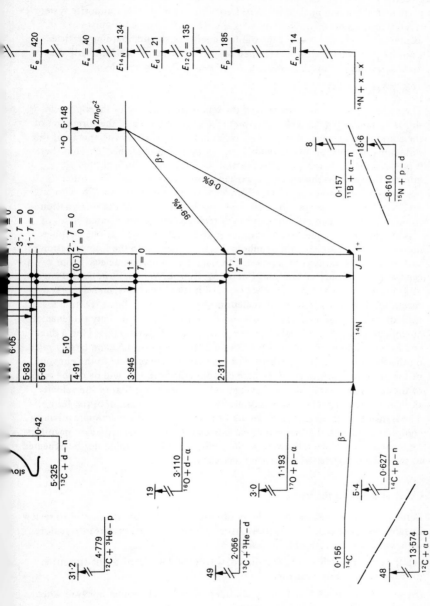

Figure 64 Energy-level diagram of ^{14}N. Note that the excitation functions (σ against bombarding energy) are plotted with centre-of-mass energy linear along the vertical axis although the figures quoted against peaks are for particle energy in the laboratory frame

particle approaches close enough to the target nucleus for nuclear forces to be effective, states of the system resulting from the incorporation of the bombarding particle in the target nucleus are very frequently formed. The composite system is referred to as *the compound nucleus* and when it is formed it must have an excitation energy in excess of the binding energy of the bombarding particle. Consider for example the well-studied reaction in which ^{10}B is bombarded with α-particles and protons are emitted. We represent this reaction by

$$^{10}B + {}^4He \rightarrow {}^{13}C + {}^1H.$$

The yield of protons as a function of the bombarding energy of the α-particle shows certain peaks. These are interpreted as corresponding to the formation of the compound nucleus, in this case ^{14}N, in an unbound state. The energy of this state must be in excess of the binding energy of an α-particle in ^{14}N. That there is physical reality in regarding the reaction as taking place through this intermediate state is shown by the existence of similar resonances in other reactions, for example ^{12}C + ^{2}H, which lead to unbound states of ^{14}N with precisely the same energies. In Figure 64 is gathered the wealth of information from a variety of such reactions and their relationships to the excited states of ^{14}N are shown. The question of the competition between the various possible modes of de-excitation of these unbound states is central to the interpretation of such nuclear reactions as they proceed through a compound nucleus. From a general study of nuclear reactions (see W. M. Gibson, *Nuclear Reactions*, Penguin, 1971) it is clear that there are other types of nuclear reactions, namely *direct reactions*, in which a compound nucleus is not created. In a direct reaction a nucleon or group of nucleons is observed to be emitted without any evidence of an unbound state being involved in the process. The distinction between a direct and a compound nuclear reaction becomes difficult to maintain and largely rests on the time elapsing from the arrival of the bombarding particle to the emission of the products. This time, in the case of a direct interaction, will be of the order of the time taken for a nucleon having a velocity one tenth that of the velocity of light to cross a nucleus, which typically will have a dimension of ten fermis. The transit time is thus a few times 10^{-22} s. In the case of a reaction involving a compound nucleus, the time would be expected to be several orders of magnitude longer than this. Unless this were so the compound nucleus could not be expected to show the properties of a quasistationary state.

10.3 Classical radiation theory

10.3.1 Before turning to the quantum theory of radiative transitions it is useful to review certain aspects of classical radiation theory which give some insight into features with which we shall later be concerned. These are

(a) the rate at which energy is radiated, which will be related to the transition probability of quantum theory,

(b) angular momentum and parity considerations, which will be involved when we come to consider selection rules, and

(c) the damping of the radiating system, which gives rise to finite widths associated with the resonant states.

Classically the nucleus is equivalent to a system of moving point charges and magnetic dipoles. In section 7.5 it was shown that the electrostatic potential arising from the most general stationary charge distribution could be conveniently treated as the superposition of the potentials of charge multipoles. Similar considerations, when applied to the time-dependent electric and magnetic radiation fields, show that these may be treated as the superposition of radiation fields from multipoles which, in the case of radiation, will be either electric or magnetic multipoles which vary with time (A. E. S. Green, *Nuclear Physics*, McGraw-Hill, 1955). We shall later see that, in practical cases, the main contribution to the radiation field comes from the lowest-order multipoles. In the static case there was seen to be a monopole contribution corresponding to an equivalent point charge. As we shall be dealing with nuclei of fixed charge, this term has no time variation and hence the lowest-order multipole with which we are here concerned is the dipole. We shall now consider some properties of the radiation field associated classically with an electric dipole. We recall that the nucleus has no static electric dipole moment (see section 7.6) in the ground state or in an excited state which has a definite parity. However we shall later see that there is a quantity equivalent to a time-varying electric dipole moment involved in the transition between stationary states. This gives rise to radiation analogous to the radiation from a classical oscillating electric dipole.

10.3.2   *Classical electric dipole*

Radiation rate. We consider a charge q having a displacement

$z = a_0 \sin \omega t,$

and therefore having an oscillating dipole moment

$p = p_0 \sin \omega t.$

The instantaneous rate of radiation of energy per unit solid angle at an angle θ, as in Figure 65, is by classical radiation theory (F. K. Richtmeyer and E. H. Kennard, *Introduction to Modern Physics*, McGraw-Hill, 1954)

$$\frac{q^2 f^2}{4\pi c^3} \sin^2 \theta,$$

where f is the acceleration of the particle, and its velocity is much less than c. Note that there is no radiation emitted parallel to the axis of vibration.

We now write

$f = -\omega^2 a_0 \sin \omega t,$

and integrate over all angles to find that dW/dt, the rate at which energy is lost to the dipole oscillator by virtue of radiation, is given by

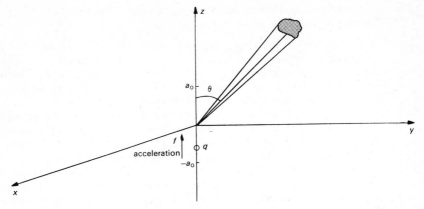

Figure 65 Oscillating dipole consisting of a charge q performing simple harmonic oscillations about the origin along the z-axis

$$dW/dt = \frac{2}{3} \frac{q^2}{c^3} \omega^4 a_0^2 \sin^2 \omega t.$$

It follows that the energy radiated, per cycle of the oscillation, by the dipole is

$$W = \frac{2q^2 \omega^4 a_0^2}{3 \quad c^3} \int_0^{2\pi/\omega} \sin^2 \omega t \, dt$$

$$= \frac{2\pi}{3} \frac{p_0^2 \, \omega^3}{c^3}.$$ **10.1**

The duration of one cycle is $2\pi/\omega$ and therefore the average rate of radiation is

$$\frac{1}{3} \frac{p_0^2 \, \omega^4}{c^3}.$$

Hence the time taken to radiate an amount of energy equal to that of one quantum, namely, $\hbar \omega$, will be

$$\frac{\hbar \, 3c^3}{p_0^2 \, \omega^3}.$$

This expression will be the classical analogue of the mean life in the quantum treatment. Consequently the transition probability λ will correspond (see section 2.3) to the reciprocal of this expression, namely

$$\frac{p_0^2}{3\hbar} \left(\frac{\omega}{c}\right)^3.$$

Angular momentum and parity. We noted above that there is no radiation of energy along the z-direction by a dipole lying perpendicular to the xy plane. Retaining the z-axis as the direction of propagation we consider now a radiating system consisting of two electric dipoles, one along the x-axis and one along the y-axis, as in Figure 66, and we suppose there to be a phase difference of $\frac{1}{2}\pi$ between them. This system is equivalent to a single circling charge, the direction of rotation deciding which dipole is leading in phase. At a distant point on the z-axis there will be an oscillating electric field vector E_x parallel to the x-axis associated with the dipole along the x-axis and a similar vector E_y parallel to the y-axis associated with the other dipole. E_x and E_y will, each in synchronism with its parallel dipole, be $\frac{1}{2}\pi$ out of phase and thus give a resultant field vector **E**, which will be of constant amplitude and will rotate with constant angular velocity. Thus the radiation along the z-axis is circularly polarized.

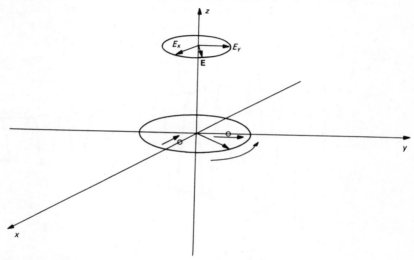

Figure 66 Circularly polarized electric field arising from a charge circulating in the xy plane. The relative phase of **E** and the particle motion will depend on distance along the z-axis

If this radiation were truly transverse (i.e. **E** and **H** perpendicular to each other and to the radius vector) the flow of energy, which is along the direction of the Poynting vector and therefore perpendicular to **E** and **H**, would be radial. However an exact analysis shows that this is not the situation at any finite distance (J. M. Blatt and V. F. Weisskopf, *Theoretical Nuclear Physics*, Wiley, 1963). The Poynting vector is not truly radial. There is thus a 'swirling' of energy about the direction of propagation and associated with this an angular momentum about that direction. This angular momentum will change sign with change in the direction of rotation of the charge and the consequent change in the sense of the circular polarization.

If the parity transformation (substitution of $-x$ for x, $-y$ for y and $-z$ for z) is applied to the system, it amounts to a diametrical displacement of the rotating charge and hence to a change of π in the phase of both equivalent dipoles. It therefore amounts to changing the sign of **E** and **H**. There is thus a negative parity associated with this electric dipole radiation.

We note that had we been dealing with a magnetic dipole, which is equivalent to a continuous current rather than a circulating localized charge, the system would not have been altered by the parity transformation. There is thus a positive parity associated with magnetic dipole radiation.

Radiation damping. Reverting to the simple electric dipole of Figure 65, we note that, as a simple harmonic oscillator, it will have an energy content equal to the potential energy when the amplitude is a maximum, i.e. proportional to a_0^2. We now assume that the energy radiated in one period of the oscillation, given by equation **10.1**, is very small compared to the total instantaneous energy contained

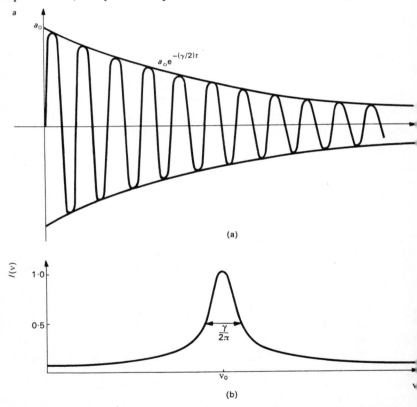

Figure 67 (a) Amplitude variation with time in damped oscillator. (b) Frequency spectrum of damped oscillator

in the oscillator. The energy radiated, being proportional to a_0^2, is thus proportional to the energy stored in the system. Thus the stored energy will decay exponentially with time. If then we write the energy W of the system as

$$W = W_0 e^{-\gamma t},$$

the amplitude of the oscillation will take the form

$$a = a_0 e^{-(\gamma/2)t}.$$

A plot of the amplitude is shown in Figure 67(a). The technique of Fourier analysis may now be applied to this exponentially decaying function to show that it can be constructed from a superposition of undamped harmonic functions having amplitudes dependent on their frequency and given by the spectral function

$$I(\nu) = \frac{\text{constant}}{(\nu - \nu_0)^2 + \frac{1}{4}(\gamma/2\pi)^2}.$$

This spectral function is illustrated in Figure 67(b). We note that

$$I\left[\nu_0 \pm \frac{\gamma}{4\pi}\right] = \frac{1}{2}I(\nu_0),$$

and therefore that the full width at half-maximum of this function is $\gamma/2\pi$. We thus see that, in so far as there is energy radiated, the resonant system is damped and the resonance is no longer infinitely sharp. The half-width is intimately connected with the rate of radiation of energy.

10.4 Quantum theory of radiative transitions

An exact treatment of radiation transitions in terms of quantum mechanics is of a difficulty beyond the scope of this text. It is to be found in an introductory form in R. M. Eisberg, *Fundamentals of Modern Physics*, Wiley, 1961, and in advanced form in J. M. Blatt and V. F. Weisskopf, *Theoretical Nuclear Physics*, Wiley, 1963. We shall here be content to give some of the physical background to the theory and to understand and apply the final results.

The transitions with which we are concerned are spontaneous transitions of an atomic or nuclear system from a state, which we shall characterize by a, to a state of lower energy, characterized by b. These transitions are to be distinguished from *induced* transitions, which are between the same states but stimulated by the presence of an electromagnetic field having a frequency given by $h\nu_{ba} = E_a - E_b$. However, from a consideration of the thermodynamic equilibrium in a closed volume within which a set of atoms or nuclei are emitting and absorbing radiation, the transition rate for spontaneous emission can be related to the transition rate for induced emission and also to the transition rate for absorption. Attention is therefore directed to the problem of the behaviour of the quantum system when it is exposed to an electromagnetic wave, and as a result absorbs or emits energy.

This problem is dealt with by *perturbation theory*. The physical basis of this theory can be described as follows. The wave function of the system in all circumstances can be expressed as a sum of terms according to the equation

$$\Psi = \sum a_n e^{-iE_n t/\hbar} \psi_n .$$

If we assume the system to be in the lower eigenstate b initially, then all the coefficients a_n are zero except a_b and we have

$$\Psi = a_b e^{-iE_b t/\hbar} \psi_b. \qquad \textbf{10.2}$$

When a perturbing field is applied, the effect is to mix into the wave function a contribution from the wave function of the eigenstate a, so that now we have

$$\Psi = a_b e^{-iE_b t/\hbar} \psi_b + a_a e^{-iE_a t/\hbar} \psi_a. \qquad \textbf{10.3}$$

Eventually the first term in this expression is reduced to zero and finally we are left with the wave function

$$\Psi = a_a e^{-iE_a t/\hbar} \psi_a \qquad \textbf{10.4}$$

and the system is then in the eigenstate a.

Now, whereas the time-dependent factor in the product $\Psi^*\Psi$ becomes unity for the wave functions given by equations **10.2** and **10.4**, in the case of the function given by **10.3** we have

$$\Psi^*\Psi = a_b^* a_b \psi_b^* \psi_b + a_a^* a_a \psi_a^* \psi_a + a_a^* a_b e^{i(E_a - E_b)t/\hbar} \psi_a^* \psi_b +$$

$$+ a_b^* a_a e^{-i(E_a - E_b)t/\hbar} \psi_b^* \psi_a.$$

We note that the product $\Psi^*\Psi$ gives the probability density for the system. When the system is in one or other of its eigenstates we see that the probability density, and hence the charge distribution, is independent of time. When, however, the system is in the state described by equation **10.3**, the probability density, and hence the charge distribution, has terms which oscillate with time. We note that the frequency of oscillation of these terms is $(E_a - E_b)/\hbar$, which is precisely the frequency of the photon absorbed or emitted in the transition.

Perturbation theory, as was noted earlier in section 4.9, relates the probability of the transition taking place to the perturbation of the system through *matrix elements*. These matrix elements are constructed from the wave functions of the initial and final states together with an operator which represents the energy arising from the interaction of the quantum system with the perturbing field. The formal definition of the matrix element in the present case will be

$$v_{ab} = \int \psi_a^* V \psi_b \, d\tau,$$

where V is the operator corresponding to the energy of the component charges of the system in the perturbing electric field of the photon.

In the nuclear context we are contemplating photons of energy up to about ten million electronvolts and having wavelengths of 130 fm and upwards. The diameter of a nucleus at the most is about 15 fm. Thus the oscillating electric field **E** produced by the photon will, to a first approximation, have a constant value throughout the nuclear volume at any instant. We consider the field **E** in terms of its components E_x, E_y, E_z. If we choose the centre of the system as the reference point from which potential energy is to be measured, then the electrostatic energy, considering for the present only the x-direction, may be written as

$$V_x = -E_x \sum_r e_r x_r.$$

There will be similar expressions for V_y and V_z. When we now construct the matrix elements according to the formula above, we find occurring integrals of the form

$$\int \psi_a^* \sum_r e_r x_r \psi_b \, d\tau,$$

which from its resemblance to

$$\int \psi_b^* \sum_r e_r x_r \psi_b \, d\tau,$$

the quantum-mechanical analogue of the classical electric dipole moment, we refer to as the matrix element of the x-component of the electric dipole moment, and denote by μ_{xab}. The transition rate for absorption is then found to be

$$\frac{E_x^2}{\hbar^2} (\mu_{xab}^* \mu_{xab} + \mu_{yab}^* \mu_{yab} + \mu_{zab}^* \mu_{zab}),$$

on the assumption that the radiation is not polarized, so that $E_x = E_y = E_z$. When this expression is substituted into the equation relating the spontaneous transition rate S_{ab} to the absorption rate, the dependence on the electric field disappears and it is found that

$$S_{ab} = \frac{32\pi^3 \nu_{ab}^3}{3\hbar c^3} (\mu_{xab}^* \mu_{xab} + \mu_{yab}^* \mu_{yab} + \mu_{zab}^* \mu_{zab}). \qquad \textbf{10.5}$$

A comparison of equation **10.5** with the expression derived in section 10.3.2 for a classical system shows that the result in the quantum case is in agreement with that for a classical oscillating dipole, whose maximum dipole moment p_0 is given by

$$p_0^2 = 4(\mu_{xab}^* \mu_{xab} + \mu_{yab}^* \mu_{yab} + \mu_{zab}^* \mu_{zab}).$$

10.5 Selection rules for radiative transitions

10.5.1 *Angular momentum (electric dipole)*

The transition probability is seen from equation **10.5** to be finite provided at least one of the dipole matrix elements has a non-zero value. If all the matrix elements are zero, then clearly the transition probability is zero and the transition is said to be forbidden. We now examine the matrix elements to establish the conditions under which they will have non-zero values, and the transition will consequently be allowed.

We are concerned basically with

$$\int \psi_a^* x \psi_b \, d\tau$$

and corresponding integrals involving y and z. To evaluate the integrals requires a knowledge of the wave functions of the two states. We proceed to consider the special case where these wave functions are of the form found in section 6.4 for a single particle in a spherical potential well. The results we shall arrive at are in fact of general validity.

We express the volume integral in spherical coordinates, substitute

$$x = r \sin \theta \cos \phi,$$

represent the wave function as the product $R(r)\Theta(\theta)\Phi(\phi)$ and then have

$$\int_0^\infty R_a^* R_b r^3 \, dr \int_0^\pi \Theta_a^* \Theta_b \sin^2 \theta \, d\theta \int_0^{2\pi} \Phi_a^* \Phi_b \cos \phi \, d\phi.$$

Taking the third (or azimuthal) integral and substituting the solution $\Phi(\phi) = A e^{i(m\phi + B)}$, we have for this integral, apart from constant factors,

$$\int_0^{2\pi} e^{-im_a \phi} e^{im_b \phi} \cos \phi \, d\phi = \tfrac{1}{2} \int_0^{2\pi} [e^{-i(m_a - m_b + 1)\phi} + e^{-i(m_a - m_b - 1)\phi}] d\phi,$$

where m_a and m_b are the magnetic quantum numbers of the two states. Now the expression

$$\int_0^{2\pi} e^{iP\phi} \, d\phi,$$

where P is integral, is zero unless $P = 0$. Hence the azimuthal integral is zero unless $m_a - m_b = \pm 1$, in which case one of the two terms in the above expression will remain finite. By examining the corresponding integral for y it can readily be shown that exactly the same conditions apply.

In the case of the z-coordinate, z is independent of ϕ, and the azimuthal integral reduces to

$$\int_0^{2\pi} e^{-i(m_a - m_b)\phi} \, d\phi.$$

This integral is zero unless $m_a = m_b$.

We arrive at the following conclusion, which is of general validity, namely that the dipole transition may take place only if the difference in magnetic quantum numbers of the two states, Δm, is ± 1 or 0. If $\Delta m = \pm 1$, then $\mu_{zab} = 0$ and, if the transition takes place, the radiation will correspond to that from the classical system of Figure 66. If $\Delta m = 0$, then $\mu_{xab} = \mu_{yab} = 0$ and in this case, if the transition takes place, the radiation will be as for the classical system of Figure 65. We later give further consideration to these two cases.

If, still in terms of the wave functions of a single particle in a potential well, the 'polar' integral

$$\int \Theta_a^* \, \Theta_b \, \sin^2\theta \, d\theta,$$

which will involve an orbital quantum number l as well as m for each state, is evaluated, it is found that, in addition to the Δm requirements above, Δl must be ± 1; otherwise the polar integral is zero.

Up to this point we have ignored the intrinsic spin of the single particle. We have to take the spin into account in order to consider the selection rules applicable to the total angular momentum of the particle, j, where $j = l \pm s$. These selection rules are to be found by using a more complete wave function taking into account the spin–orbit interaction. Here we are content to quote the results of such an analysis, namely that for electric dipole transitions $\Delta j = \pm 1$, as for Δl, but that the further possibility of $\Delta j = 0$ also exists. In this latter case Δl is compensated by a change in the spin quantum numbers, a 'spin flip' taking place.

We can now translate the selection rules into terms of the initial and final spin quantum number of two nuclear states. In terms of the single-particle shell model, I, the nuclear spin, is to be equated with j for an odd-A nucleus, and the angular-momentum selection rule for electric dipole radiation becomes $\Delta I = \pm 1$ or 0. The validity of this rule is not limited to states which are describable in terms of the single-particle shell model but is of general validity. Further, it applies also to the states of even-A nuclei, with the proviso that a transition from a state having $I_a = 0$ to a state having $I_b = 0$ (which is a particular case of $\Delta I = 0$) is forbidden. This forbiddenness is understandable since in this case both initial and final states are states of spherical symmetry.

We can represent the angular-momentum selection rules graphically. In Figure 68 the relationships which have to exist between initial- and final-state values of l and m for a single particle are shown for electric dipole transitions.

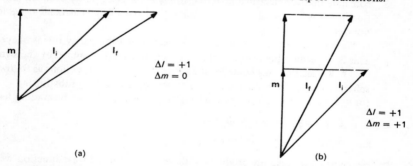

Figure 68 Vector diagram illustrating the selection rules for electric dipole transitions

We can maintain, in considering these diagrams, the picture of a photon having an associated spin 1. In the case of Figure 68(b) we recall that for $\Delta m = \pm 1$, $\mu_{zab} = 0$. We therefore have a situation analogous to the classical case illustrated in Figure 66. The radiation emitted along the z-axis is circularly polarized and carries angular momentum directed along its direction of propagation. This angular momentum carried by the photon is of course exactly equal to the angular momentum lost by the nuclear system. The radiation emitted in the xy plane is plane polarized and carries angular momentum at right angles to its direction of propagation, i.e. parallel to the z-axis, as before. At oblique angles the polarization is elliptic, i.e. a combination of the two cases discussed above; the associated angular momentum is still parallel to the z-axis. In the case of the situation represented in Figure 68(a), we recall that for $\Delta m = 0$, $\mu_{xab} = \mu_{yab} = 0$ and therefore the corresponding classical situation is that of a dipole oscillating along the z-axis as in Figure 65. In this case there is no emission in the z-direction and the radiation in all other directions is plane polarized with the associated angular momentum lying parallel to the xy plane.

In the case of the nucleus, the diagrams of Figure 68 are still valid if we replace $\mathbf{l}_a$ and $\mathbf{l}_b$ by $\mathbf{I}_a$ and $\mathbf{I}_b$. However in the case of the nucleus there is the additional possibility that $\Delta I = 0$. In this event $\mathbf{I}_a$ and $\mathbf{I}_b$ are to be represented as making equal angles with the z-axis and hence being two generators of a cone with axis Oz such that $I_a = I_b + 1$. This vector representation is geometrically possible for all values of I_a and I_b except $I_a = I_b = 0$, which, as we saw above, is not an allowed transition.

10.5.2 *Parity selection rules for electric dipole transitions*

If we apply the parity transformation to the function $\psi_a^* x \psi_b$ we find that there is a change of sign if ψ_a and ψ_b have the same parity and no change of sign if ψ_a and ψ_b have opposite parity. This behaviour arises, of course, from the odd parity of the factor x. It follows therefore that the integration over all space of this function leads to a finite result only if ψ_a and ψ_b have opposite parity. Hence we conclude that for electric dipole transitions a change of parity between initial and final states is a necessary condition for at least one of the matrix elements to be non-zero.

We note that the condition $\Delta l = \pm 1$ ensures that this parity rule is obeyed. This is so because the parity of the wave functions is determined by the sign of $(-1)^l$. We see that this holds for the single-particle wave functions as follows. In spherical coordinates, the parity transformation amounts to substituting $\pi - \theta$ for θ, $\pi + \phi$ for ϕ and leaving r unchanged. These angular transformations lead to

$$\Phi_m(\pi + \phi) = (-1)^{|m|} \Phi_m(\phi)$$

and $\Theta(\pi - \theta) = (-1)^{l + |m|} \Theta(\theta)$.

Hence when we take the products of these with the radial wave function we have

$$\psi(r, \pi - \theta, \pi - \phi) = (-1)^l \psi(r, \theta, \phi).$$

It therefore follows that two states differing by one in l-value, as do the states involved in an electric dipole transition, necessarily have opposite parity.

Selection rules for radiative transitions in the general case

Many radiative transitions are observed to take place between states whose spins and parities do not satisfy the above selection rules. Only in the case of $I_a = 0 \rightarrow I_b = 0$ is the transition found to be absolutely forbidden. To understand why the transitions still can take place, despite the selection rules not being obeyed, we have to have regard to the assumption made above concerning the nature of the interaction between the nucleus and the electrostatic field. It was assumed that the electrostatic field acted uniformly throughout the nuclear volume. This uniformity is only true to a first approximation and arises from the nuclear dimension being very much smaller than the wavelength of the radiation. It amounts to taking the first term in the series expansion of the harmonic function representing the space dependence of the electromagnetic wave. There will be further terms involving higher powers of x/λ which we neglected. If these terms are included, then the matrix element has to be expressed as a sum of terms of the form

$$\int \psi_a^* x^L \psi_b \, d\tau,$$

where L is now $1, 2, 3, \ldots$. The terms in the series get progressively smaller as L increases. However, if, because of a contravention of the selection rules derived above, the term corresponding to $L = 1$ becomes zero, then the transition probability is dominated by the term corresponding to $L = 2$. This integral is termed the electric-quadrupole, or E2, matrix element.

There is the further consideration that no allowance has been made for the interaction of the magnetic field of the wave with the currents and magnetic dipoles involved in the system. Allowance for this interaction introduces magnetic-multipole matrix elements corresponding to magnetic multipole transitions, which we denote by M1, M2, etc.

We consider first the selection rules for E2 transitions. These are arrived at by a consideration of

$$\psi_a^* x^2 \psi_b \, d\tau.$$

Applying the same simple argument as we applied to the electric dipole (i.e. E1) transition above, we see that, since x^2 has positive parity, the initial and final states must be of the same parity, otherwise the E2 matrix element will be zero. The arguments made above with respect to the single-particle wave functions can be repeated for the E2 matrix element and the selection rule arrived at is that $\Delta I = \pm 2, \pm 1$ or 0, with again $I_a = 0 \rightarrow I_b = 0$ being forbidden.

We consider now M1 transitions. We noted above that a magnetic dipole behaves in the opposite manner to an electric dipole under parity transformation. It follows that the parity selection rule for M1 transitions is that the states involved be of similar parity. The angular-momentum selection rules are found in the case of M1 transitions to be the same as for E1 transitions.

In the case of an L-pole transition, the photon can be considered to take away an angular momentum $L\hbar$. According to the conservation of angular momentum, the vectors I_a, I_b and L must therefore form the sides of a triangle. This limits the possible values of L, for given I_a and I_b values, to those satisfying the conditions

$$|I_a - I_b| \leqslant L \leqslant I_a + I_b.$$

The electric or magnetic character of the transition is determined by the relative parities of the initial and final states, there being no change of parity for electric transitions of even L-values and for magnetic transitions of odd values.

10.6   Transition rates

The transition rate given by equation **10.5** we now see to be valid for E1 transitions only. When we pass to higher multipoles we observed above that we are incorporating additional terms from the expansion of a space factor $\exp(i2\pi x/\lambda)$ in the expression for the field of the electromagnetic wave. As therefore we pass from one multipole order to the next we have to expect an additional factor x/λ to appear in the interaction energy involved in the matrix element. Not only does this introduce the next highest nuclear multipole into our considerations by increasing by one the power of x, but it introduces a factor proportional to $1/\lambda$ (and hence proportional to ν_{ab}/c) into the matrix element. The transition rate, being proportional to the square of the matrix element, in the general case is therefore expected to be proportional to $(\nu_{ab}/c)^{2L+1}$.

A detailed treatment of the general case, given by J. M. Blatt and V. F. Weisskopf *Theoretical Nuclear Physics*, Wiley, 1963, shows that the result may be written as

$$\lambda = \frac{8\pi(L+1)}{L[(2L+1)!!]^2} \frac{1}{\hbar} \left[\frac{2\pi\nu_{ab}}{c} \right]^{2L+1} B_{ab}(L),$$

where $(2L+1)!!$ represents the product $1 \times 3 \times 5 \times \ldots \times (2L+1)$. When λ, here the transition probability, is written in this way, all of the dependence of the transition rate on the nuclear wave functions is contained in $B_{ab}(L)$, which is termed the *reduced transition probability*. When comparing transition rates for transitions of the same character, for different pairs of states, complications arising from the different transition energies are clearly avoided by comparing the values of $B_{ab}(L)$ rather than comparing the values of the transition probability.

In the case of magnetic transitions, current density plays the role played in electric transitions by charge density. This introduces a factor v/c in the matrix element, and hence a factor $(v/c)^2$ in the transition probability for magnetic transitions as compared to electric transitions, where v is the nucleon velocity and is approximately $\frac{1}{10}c$. The existence of nuclear magnetic moments complicates the issue but does not prevent the magnetic transition rate being significantly smaller than an electric transition of the same multipole order, due allowance having been made for the energy dependence.

The calculation of exact values for transition probabilities requires a detailed knowledge of the nuclear wave functions of the states; usually this is not available. In fact an excellent test of any proposed wave functions is to use them to predict transition rates which may then be compared with measured values. For many purposes, however, it is valuable to compare measured transition rates with those calculated on the basis of the extreme single-particle model. On the assumption of this model, namely that the transition involves a change of state of one nucleon only, Weisskopf (1951) arrived at the following estimates for reduced transition probabilities in the case of electric and magnetic multipoles.

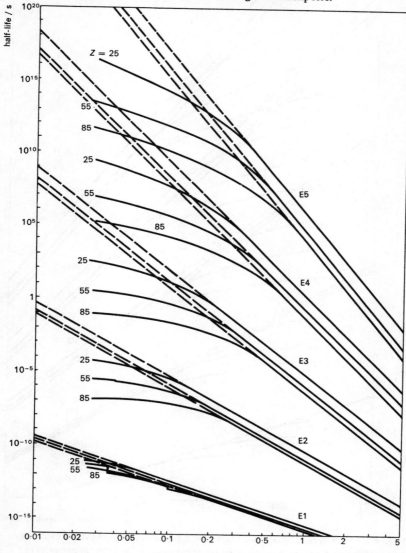

Figure 69 Half-lives for electric multipole γ-ray emission (Weisskopf estimates) with (solid lines) and without (dashed lines) correction for internal conversion

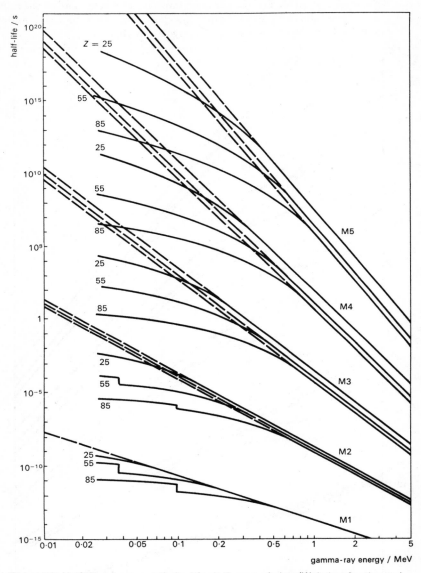

Figure 70 Half-lives for magnetic multipole γ-ray emission (Weisskopf estimates) with (solid lines) and without (dashed lines) correction for internal conversion

$$B(EL) = \frac{e^2}{4\pi} \left[\frac{3R^L}{L+3} \right]^2 ,$$

$$B(ML) = 10 \left[\frac{\hbar}{McR} \right]^2 B(EL),$$

where R is the nuclear radius and M is the proton mass. We may further introduce $R = R_0 A^{\frac{1}{3}}$ and replace the R-dependence by an A-dependence.

These estimates can now be converted into half-life for the excited state. The results of such a conversion are shown in Figures 69 and 70 plotted against the energy of the transition. It is seen that the lowest multipole order permitted by the selection rules is favoured and that electric transitions always proceed faster than magnetic transitions of the same multipole order.

10.7 **Level widths**

We now consider a transition from an excited state to the ground state, which we assume stable. Just as in the classical theory, as we saw in section 10.3.2, a radiating system is damped and emits a range of frequencies, so in the quantum case we have to assume that the energy level corresponding to an excited state has an energy width which we represent by Γ. We can relate Γ to the mean life τ of the state by Heisenberg's uncertainty principle, which gives

$$\Gamma\tau = \hbar.$$

As in the classical analogue, the energy of the radiation emitted will not, strictly speaking, be monochromatic but will have a spectrum $I(E)$ given by

$$I(E) = \frac{\text{constant}}{(E - E_0)^2 + \frac{1}{4}\Gamma^2},$$

where $E_0 = E_a - E_b$.

If the final state is not stable but itself has a finite decay probability, then it also will have a level width and Γ will be the sum of the widths of the two levels.

Since $\lambda = 1/\tau$ and $\Gamma\tau = \hbar$, the values of transition probabilities quoted in section 10.6 can be readily translated into level widths. In the case of electric transitions the first few level widths are as follows (E_0 in MeV; $R = 1 \cdot 2 \times A^{\frac{1}{3}}$ fm)

$$\Gamma(E1) = 6 \cdot 8 \times 10^{-2} A^{\frac{2}{3}} E_0^3 \, \text{eV},$$

$$\Gamma(E2) = 4 \cdot 9 \times 10^{-8} A^{\frac{4}{3}} E_0^5 \, \text{eV},$$

$$\Gamma(E3) = 2 \cdot 3 \times 10^{-14} A^2 E_0^7 \text{eV}.$$

In the case of magnetic transitions the results are

$$\Gamma(M1) = 2 \cdot 1 \times 10^{-2} E_0^3 \, \text{eV},$$

$$\Gamma(M2) = 1 \cdot 5 \times 10^{-8} A^{\frac{2}{3}} E_0^5 \, \text{eV},$$

$$\Gamma(M3) = 6 \cdot 8 \times 10^{-15} A^{\frac{4}{3}} E_0^7 \, \text{eV}.$$

The resonances are thus seen to be very sharp, i.e. the energy width of the level is a very small fraction of the γ-ray energy.

10.8   **Internal conversion**

In the above discussion we have proceeded on the assumption that the system is isolated except for its interaction with the electromagnetic field. This in fact is not strictly correct either for atoms or for nuclei. In normal circumstances atoms are frequently in collision with other atoms and in their collisions energy-transfer processes are involved. As a consequence, there will always be a probability of de-excitation of an excited state by a collision in competition with de-excitation by radiation. Normally this competition is such, in the atomic context, as to suppress all radiative transitions except that of the electric dipole.

Nuclei, on the other hand, are isolated very effectively from each other in ordinary matter by the effect of their screen of orbital electrons. However, in the nuclear case, the very existence of orbital electrons provides a method of de-excitation which competes with radiative transitions. In the neighbourhood of the nucleus the field of the nuclear multipole will act on any electron in that region of space and can communicate the full transition energy to the electron. This energy will usually be much greater than the electron binding energy and consequently the electron will be ejected from the atom. This process is rather misleadingly referred to as *internal conversion*. It must not be thought of as the emission by the nucleus of a photon which is subsequently absorbed by the atomic structure of the same atom. This is a possible process but is much less likely than internal conversion. Internal conversion as a process is distinct from, and competes with, photon emission. The energy is communicated directly to the emitted electron, not by the intervention of an electromagnetic wave.

The competition with radiative transitions is clearly brought out by defining the *conversion coefficient* α as

$$\alpha = \frac{N_e}{N_\gamma},$$

where N_e and N_γ are the experimentally measured numbers of electrons and photons emitted in the same time interval from the same sample. On this definition α can take values from zero to infinity.

The energy of the emitted electron will be given by

$$E_e = E_0 - E_B,$$

where E_0 is the transition energy and E_B the electron-binding energy. In so far as electrons from different shells have finite probability densities close to the nucleus, electrons may be emitted from L, M, . . . shells as well as from the K-shell. These electrons will be experimentally distinguishable because of the difference in E_B. Experimentally we can therefore determine $\alpha_K, \alpha_L, \ldots$, where

$$\alpha_K = \frac{N_K}{N_\gamma}$$

is defined in terms of K-electrons only. It follows from these definitions that

$$\alpha = \alpha_K + \alpha_L + \alpha_M + \ldots$$

If now we introduce λ_e to represent the probability that an electron be emitted and λ_γ the probability that a photon be emitted we have

$$\alpha = \frac{\lambda_e}{\lambda_\gamma}$$

directly from the definition of α. Internal conversion and radiative transitions are assumed to proceed as independent processes and hence the total transition probability λ is given by

$$\lambda = \lambda_e + \lambda_\gamma$$

On this view the lifetime of the state, which is proportional to the reciprocal of λ, will be shortened by the existence of the phenomenon of internal conversion. Support for this view comes from the discovery by Bainbridge and others (1957) that the lifetime of a state in ^{99}Tc was altered by about 0·3 per cent by a change in the chemical composition of the sample. This effect can be entirely accounted for by the difference in the electronic wave functions of the two chemical compounds concerned.

λ_e can be developed, as α was above, in terms of conversion probabilities involving electrons from the different shells, giving $\lambda_e = \lambda_K + \lambda_L + \ldots$

The calculation of conversion probabilities is in principle straightforward. We start with an initial state consisting of an excited nucleus and a bound electron, and end in a final state with a de-excited nucleus and an electron in a state in the unbound continuum. The transition rate has then to be found from the square of the matrix elements connecting the two states and the density of final states (the so-called *golden rule* which we quoted in section 4.9). The mathematical details are of formidable difficulty even for relatively simple transitions (see E. Segrè, *Nuclei and Particles*, Benjamin, 1964). The results of the mathematical analysis show that the dependence of λ_e on the nuclear multipole moments is exactly the same as the dependence in the case of λ_γ. As a consequence α, the internal conversion coefficient, is independent of the nuclear moments. This has made internal conversion a very important phenomenon in unravelling the character of particular nuclear transitions. α is dependent on the Z-value of the atom involved, as this controls the number of available electrons. α also depends on the transition energy E_0, on the multipolarity of the transitions and on the electric or magnetic character of the transition. These latter factors control the field strength involved in the process.

The Z-dependence of α follows a high positive power (Z^3 in the case of α_K). The E_0 dependence follows a high negative power. As a consequence, internal conversion has its greatest relative importance for high-Z nuclei and low-energy transitions. It is also in these circumstances that the dependence on the character of the multipolarity is greatest. Hence internal conversion has its most significant experimental contribution to make for low-energy transitions in heavy nuclei. Its effectiveness in providing decisive evidence of the character of the transition can be seen from Figure 71, which shows the variation to be expected in the conversion coefficient for transitions of different multipole order.

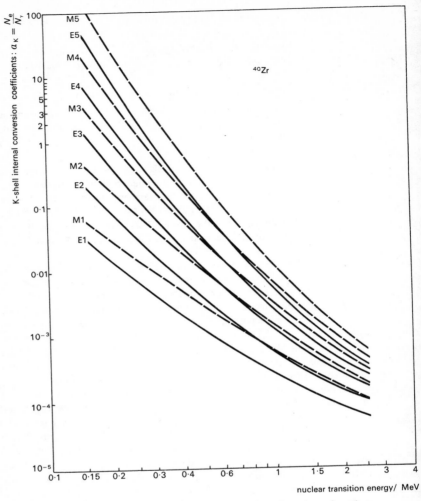

Figure 71 K-shell internal conversion coefficients as a function of nuclear transition energy for zirconium ($Z = 40$)

In addition to providing valuable information about the character of radiative transitions, the studies of internal conversion are important in two other respects. Firstly, a measurement of the energy of the emitted electron, together with a knowledge of the electron binding energy, permits a measurement of the transition energy to be made. This is a useful alternative to the methods of γ-ray energy measurements based on photoelectron production in materials external to the source. Secondly, if the transition energy is otherwise known, a measurement of the electron energy permits an accurate determination of the electron binding energy. This knowledge can then be used to determine the Z of the atom within which the transition took place. The necessary experimental accuracy is set by the fact that there is a difference of about 2·5 keV in the binding energy of K-electrons in heavy nuclei differing by one in the Z-value. The importance of this Z-determination lies in being able unambiguously to decide whether a γ-ray transition took place before or after an α- or β-decay.

Finally it is to be noted that, following internal conversion, the atom is left with a vacancy in one of the electron shells. There will therefore be the emission of an X-ray or an Auger electron. It is sometimes more convenient to detect the X-ray or Auger electron rather than the conversion electron, and this information can be used to determine N_e, due allowance being made for the X-ray fluorescence yield (i.e. the number of X-rays emitted per vacancy).

10.9   Internal pair production

According to the 'hole' theory of positrons, the excited nucleus is immersed in a sea of electrons occupying and completely filling states of negative energy which correspond to the negative square roots of $p^2c^2 + m_0^2c^4$. If the transition energy is equal to $2m_0c^2 = 1·022$ MeV, then sufficient energy is available to raise an electron from the highest of the negative-energy states to the lowest state in the positive continuum given by the positive square roots of $p^2c^2 + m_0^2c^4$. Should this occur then a stationary electron together with a vacancy in the highest

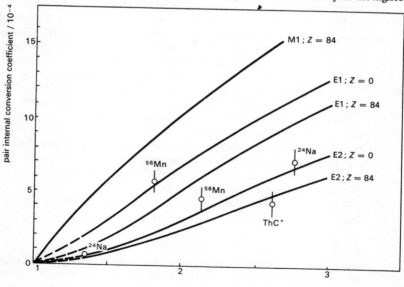

Figure 72 Internal pair conversion coefficients as a function of transition energy for nuclei of low and high Z-value

negative-energy state is produced. The vacancy constitutes a positron of zero kinetic energy. If the transition energy is greater than $2m_0 c^2$ then an electron may be lifted from a lower-lying negative-energy state across the energy gap to a higher-lying state in the continuum. This constitutes the production of an electron–positron pair, the particles sharing the energy $E_0 - 2m_0 c^2$. The availability of electrons to the nucleus does not, in this phenomenon, depend on Z. In fact the process is more likely, other things being equal, in light as compared to heavy nuclei. Also the likelihood of pair production increases with increasing transition energy. It is also favoured for low as compared to high multipole orders. In all these respects the probability of internal pair production behaves oppositely to that of internal conversion.

In Figure 72 the internal pair conversion coefficient, i.e. the ratio of pairs produced to photons emitted, is plotted as a function of energy for different multipolarities. The discrimination between multipole orders decreases with increasing transition energy. At higher energies, multipole orders can better be distinguished by the angular correlation between positron and electron, the angle between the particles being on average smaller the higher the multipole order.

10.10 0 → 0 Transitions

We saw that $0 \to 0$ transitions cannot take place by the radiation of photons. While the field outside the nuclear volume is the same for both states, and therefore no electromagnetic field is generated in the transition, there can be a change in radius and a change in field within the nuclear volume. When there is no change in parity between the two states, internal conversion is a possible process. However, the transition rate will be small because the effective volume is small compared to the volume within which internal conversion can take place for other transitions. When the transition energy exceeds $2m_0 c^2$, internal pair production is also possible, the pair being produced within the nuclear volume. There are several examples known of $0^+ \to 0^+$ transitions proceeding by internal conversion and others of similar transition proceeding by pair production.

In the case of $0^+ \to 0^-$ transitions, internal conversion and pair production, as well as single-photon radiative transitions, are forbidden. It would appear that the most likely mode of decay in such a case would be two-photon emission, or one-photon emission together with one conversion electron. Processes such as these are physically possible in the case of all transitions but have a very low probability compared to the processes discussed above.

There is no known excited state which, for lack of a possible decay mechanism, has an infinitely long life against de-excitation.

10.11 The measurement of energy-level spacing

We begin our consideration of the experimental measurements of the properties of excited states by surveying briefly the more important techniques which may be used for mapping out the energy levels. The problem is that of the accurate measurement of the energies of the emitted γ-rays. Many techniques are available for γ-ray energy measurements and we proceed to outline the most important of these.

The use of a crystal lattice as a diffraction grating is well established in X-ray technology and provides an accurate way of determining the wavelength of the X-ray. For reflection from parallel planes of atoms which have a separation d, the Bragg angle at which reinforcement occurs is given by

$$\lambda = 2d \sin \theta,$$

where θ is the angle between the X-ray beam and the reflecting surface. The extent to which this technique can be extended in the direction of shorter wavelengths, i.e. higher quantum energies, is governed by the extent to which the crystal reflecting power falls off with increasing quantum energy and the extent to which the Bragg angle gets impractically small.

While the first of these effects has to be accepted, the second can be ameliorated by experimental ingenuity. To see the smallness of the Bragg angle involved in extending the technique to γ-ray measurements, let us consider the application of the method to the measurement of the wavelength of *annihilation radiation*. This radiation, which is emitted when positrons annihilate on electrons at rest, is a very convenient standard radiation with which to calibrate γ-ray spectrometers. When an electron of zero kinetic energy encounters a positron at rest it makes a transition down to the highest of the negative-energy states. Thus the energy of the gap, namely $2m_0 c^2$, is available for radiation. However, in order that the final linear momentum of the system be zero, as was the initial linear momentum, two oppositely directed photons have to be emitted. Each photon therefore has an energy of $m_0 c^2$, which is equal to 510·98 keV. The wavelength of this annihilation radiation is therefore

$$\lambda = \frac{hc}{m_0 c^2} = 2.43 \times 10^{-9} \text{ mm}.$$

For calcite, which has a lattice constant $d = 3.02 \times 10^{-10}$ m, the Bragg angle for annihilation radiation is $\theta = 4 \times 10^{-3}$ rad. This means that a high degree of parallelism in the incident beam and a long distance of travel following reflection are necessary to separate the reinforced reflected beam from the incident beam. In most situations encountered in γ-ray spectroscopy the radiation is emerging from a small specimen so that the geometry called for using an orthodox crystal would impose a very low solid angle for acceptance of photons, and would lead to very low efficiencies. An ingenious method in which the crystal is curved enables the Bragg angle for photons sent through the crystal to be kept constant over a larger solid angle of emission from a point source (J. W. M. Du Mond, *Experimental Nuclear Physics*, Volume 3, Wiley, 1959) and makes it more practical to screen a γ-ray detector, set to detect reflected photons from the incident beam. With a curved crystal, the annihilation radiation spectrum, wavelength 2.43×10^{-12} m, has an instrumental full width at half-maximum of 0.3×10^{-13} m, i.e. about 1 per cent of the wavelength. The accuracy of determination of the position of the peak, i.e. the absolute wavelength, is believed to be 0·04 per cent. Working under these conditions the overall detection efficiency is 5×10^{-8} counts per photon emitted from a point source.

10.11.2 *Magnetic spectrometers*

The γ-ray energy determination reduces to the measurement of the energy of an electron if advantage is taken of internal conversion or if the photons are allowed to interact with atomic electrons in a thin radiator and thus to produce photoelectrons, Compton electrons or electron-positron pairs (see Appendix C). The secondary electrons may then be sent into a uniform magnetic field H and their curvature ρ in the plane perpendicular to the field measured.

The electron momentum is then given by $30\,000H\rho$, where H is in gauss, ρ in metres and the momentum in eV/c. Since the electron kinetic energy is comparable to its rest mass in these experiments, we must use the relativistic relation

$$E_e^2 = p^2 c^2 + m_0^2 c^4$$

to find the total energy of the electron. The rest-mass energy $m_0 c^2$ may be subtracted from this to find the kinetic energy T_e. If the electron is an internal conversion electron or a photoelectron from the radiator, then we have simply to add the appropriate binding energy to T_e to arrive at the transition energy. If the electron is a Compton electron or an electron from a pair, then the energy T_e for a given value of H is no longer unique. A spectrum has to be measured and compared with theoretically predicted spectra for a range of γ-ray energies.

Energy resolution in a magnetic spectrometer can only be improved at the expense of reducing the solid angle of acceptance. A typical practical compromise, in which the electrons are bent round a semicircle from a radiator to a detector separated by a distance 2ρ, provides 1 per cent energy resolution at a photon energy of 1 MeV with an efficiency of 10^{-11} counts per photon.

10.11.3 *Proportional counters*

A proportional counter usually consists of a fine wire running coaxially along a sealed cylindrical metal tube and insulated from the tube. The wire is raised to a positive potential with respect to the tube. If any ionization takes place in the sealed volume of gas, the electrons produced are accelerated towards the wire. In the high field gradient in the neighbourhood of the wire an electron can gain enough energy between collisions to ionize the next gas molecule it encounters. In the resulting avalanche, the total charge collected by the wire can, with proper experimental design, be arranged to be proportional to the number of primary electrons produced by the ionizing particle.

The gas amplification produced in this way can amount to 10^3 without any loss of proportionality. The collected current is then passed through a high resistance and the resulting voltage pulse amplified to the height required to operate a pulse-height analyser. In the range of photon energies immediately above the K-absorption edges of the gases used to fill the counter there is a high probability that a photon passed in through a thin window will be absorbed in the gas and produce a comparatively low-energy photoelectron, which will stop in volume of the counter. In this way the counter, with a choice of a gas of high atomic number, such as xenon, for a filling, can for photons of energies of 100 keV have a resolution of 3 per cent and an efficiency of 10 per cent.

In some cases the sample may be in gaseous form and suitable for introduction as a counter gas. Then, for low-energy transitions, K and L internal conversion electrons will stop in the gas; they are detected with 100 per cent efficiency and a very accurate measurement, not only of the transition energy but of the ratio α_K/α_L is possible.

10.11.4 *Scintillation spectrometers*

Charged particles produce not only ionization along their tracks but leave a trail of excited atoms which de-excite by radiating photons in the optical range of the electromagnetic spectrum. The light output from the particle track is proportional to the energy dissipated by the particle, which will be its total kinetic energy if the particle stops within the material. Certain crystals are quite transparent to this internally produced light. When this is so, the emergent light can be collected and passed to a photomultiplier, which will produce a charge proportional to the number of photons of light falling on its photocathode. This charge can be passed through a resistance and the voltage pulse amplified. Voltage pulses produced in this way and passed to a pulse-height analyser will then reproduce the energy spectrum of the charged particles.

A scintillating crystal, in addition to the necessary optical properties, requires to have a proportion of heavy nuclei incorporated in it if it is to efficiently convert incident photons into secondary electrons and stop them within its own volume. This condition is admirably satisfied by sodium iodide, which has found a very wide application in γ-ray spectroscopy. The crystal may be several centimetres in linear dimension and large enough to be operated under conditions approaching total absorption of the incident photon. In this case, (a) when photoelectric absorption takes place, not only is the photoelectron stopped in the crystal but the subsequently emitted X-ray is also absorbed; (b) when Compton scattering takes place, not only is the scattered electron stopped but the degraded γ-ray is absorbed and; (c) when pair production takes place, not only are both particles stopped but the annihilation quanta are also absorbed in the crystal. The crystal light output is then proportional to the total energy of the photon and a single peak in the pulse distribution is observed for a single γ-ray energy. In most cases, however, all the secondary photons are not absorbed and subsidiary peaks, *photon escape peaks*, are observed in the spectrum.

In the event of the source emitting two γ-rays from a cascade within a time short compared to the resolving time of the crystal and the associated electronic apparatus (say 10^{-6} s) then it is possible for both γ-rays to produce light pulses

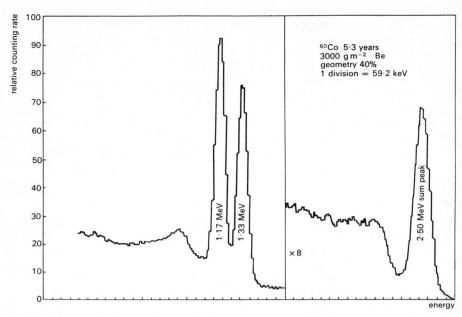

Figure 73 Energy spectrum of ^{60}Co measured with 10 cm x 10 cm NaI(Tl) crystal showing total energy peaks for the γ-ray lines at 1·17 MeV and 1·33 MeV together with the sum peak. The γ-rays were passed through a beryllium filter

which add together at the photocathode of the multiplier. There is thus the possibility of a *sum peak* in the spectrum. An indication of the attainable performance from a 10 cm x 10 cm cylindrical sodium iodide (thallium activated) crystal is given in Figure 73. The counts per emitted photon from a source can exceed 0·1 and, as is seen from the figure, the energy resolution is about 8 per cent. The energy resolution depends both on the statistical fluctuations in light emitted as a function of energy dissipated and in the variation of efficiency of light collection from different regions of the crystal.

Scintillation counters offer very high efficiencies compared to magnetic spectrometers or curved-crystal spectrometers but their resolution is considerably inferior.

10.11.5   *Solid-state detectors*

The advent of solid-state detectors has enabled γ-spectroscopy to proceed with the high resolution of magnetic spectrometers and with an efficiency approaching that of heavy scintillators. This, as we shall see, has had very important consequences in experimental nuclear physics.

The physical principle involved in solid-state detectors concerns electrical conductivity in a crystal. In the case of a perfect crystal, all the electrons are in the filled band which lies, with a band gap of about one electronvolt, below the conduction band. If two electrodes are connected to opposite faces of the crystal and a voltage difference maintained between them, there will be no flow of current. However, a charged particle passing through the crystal can promote the electrons from the filled to the conduction band. There is a consequent flow of current. The energy necessary to promote an electron and thus produce an electron–'hole' pair is on the average about three electronvolts, which is to be compared with thirty electronvolts necessary to produce an ion pair in a gas. The movement of the electron and the 'hole' through the crystal is very fast compared with the movement of electrons and positive ions in a gas counter. It is therefore to be expected, because of the larger number N of electron–hole pairs produced compared to ion pairs for a given energy dissipated, that the statistical variation, which will be proportional to $1/\sqrt{N}$ will be smaller and that this will lead to better energy resolution. At the same time, because of the high mobility of the electrons and holes, the counter can have a very fast response time.

The achievement of a practical detector based on this principle is not however a simple matter. In a real as distinct from an ideal crystal, the conduction band is never in fact completely unoccupied. Electrons are raised into it by thermal fluctuations and, usually more importantly, by the action of impurity centres in the crystal. In the semiconductors silicon and germanium, it has proved possible by 'doping' (i.e. by introducing impurities in a controlled way) to compensate for the unavoidable natural imperfection of the crystal and so to approach the behaviour of an ideal crystal. This has been achieved by drifting lithium ions moving under the influence of an applied electric field into a pure silicon or germanium crystal. Germanium has a special interest for γ-ray spectroscopy because of its high atomic number. In practice there is the complication that the crystal has to be used and stored at liquid-nitrogen temperature, otherwise its performance is seriously impaired by lithium diffusing out.

Germanium crystals with a sensitive volume of a hundred cubic centimetres are now commercially available. Their efficiency is such that the number of counts in the total absorption peak of the spectrum can be 2–3 per cent of the number of γ-rays falling on the detector for γ-rays of two million electronvolts energy. The efficiency rises to much higher values at lower energies but falls off quite rapidly at higher energies. The intrinsic time resolution of the detector can be less than 10^{-8} s.

A spectrum of ^{60}Co measured with a 40 cm^3 lithium drifted germanium detector is shown in Figure 74. Comparison of this spectrum with the sodium iodide spectrum in Figure 73 gives a clear indication of the improvement effected by the development of solid-state detectors. With their help it is now possible, even when only comparatively weak sources are available, to measure level spacing to five significant figures.

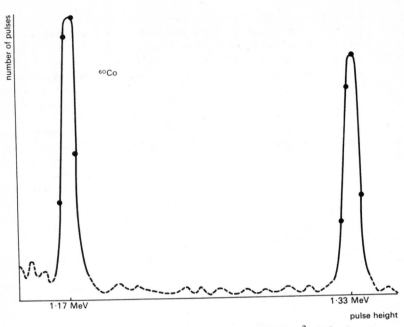

Figure 74 Energy spectrum of ^{60}Co measured with a 40 cm^3 Li–Ge detector, showing the total energy peaks for the 1·17 MeV and 1·33 MeV γ-ray lines

10.12 The measurement of lifetimes of excited states

The measurement of the lifetimes of excited states, when the transitions are of high multipole order and therefore fall into the category of isomeric transitions, as discussed in section 6.10, represents the same problem as the measurement of the lifetimes of α- and β-emitters. The numbers of γ-rays $N_γ$ emitted in equal intervals of time, shorter than or comparable with the half-life, are measured. A plot of $\ln N_γ$ against t then results in a straight line of negative slope λ, where λ is the total transition probability. Electronic timing of the intervals, with the counts being delivered to a set of scaling circuits and mechanical counters, allows this direct technique to be extended down to half-lives in the microsecond (10^{-6} s) region. Using only the front edges of pulses from scintillation detectors, or using fast pulses from solid-state detectors in conjunction with delay lines and fast coincidence circuits, half-lives down to 10^{-10} s may be measured by an extension of the direct technique. There is still a gap of many orders of magnitude between this range of lifetimes and the half-lives anticipated for lower multipole order transitions.

In certain circumstances a further range of half-lives extending down to picoseconds (10^{-12} s) can now be explored using the Doppler shift of γ-ray energy which occurs when the emitting nucleus is recoiling. This extension has proved important, as we shall see later, in the information that it has given concerning the transition probabilities of E2 transitions.

We begin discussion of this method, referred to as the *Doppler shift attenuation method* (DSAM) by making the unrealistic assumption that we are dealing with a set of nuclei all produced in excited states by bombardment with heavy projectiles, and all recoiling into vacuum in the same direction with identical velocities. Suppose a stop is arranged as shown in Figure 75, and make the further

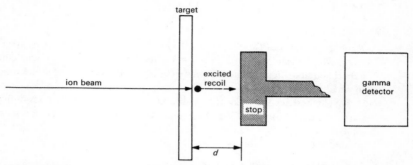

Figure 75 Schematic representation of Doppler shift attenuation measurement of lifetime of excited state

assumption that the recoils, which will be highly charged particles, are brought to rest instantaneously when they enter the stop. If the mean life τ of the excited state is short compared to the time of transit of the recoils from the target to the stop, d/v, then most of the nuclei will de-excite in flight before hitting the stop. The wavelength (and hence frequency and energy) of the radiation will be modified by the velocity of the source according to the well-known Doppler effect. If $v \ll c$, the photons emitted by the recoiling nuclei at $0°$ to their direction of travel will have an energy

$$E = E_0 \left[1 + \frac{v}{c} \right],$$

where E_0 is the transition energy. The recoils which survive the transit without de-exciting will be stationary in the stop when emission takes place. They will emit photons of the full energy E_0.

The analysis developed in Chapter 2 for radioactive decay can of course be applied directly to the de-excitation of the excited nuclei. The probability that a nucleus will *not* decay in a time t, measured from its instant of emission from the target, is given by $\exp(-\lambda t)$. Thus, the time of flight from target to stop being d/v, the ratio of nuclei de-exciting in flight will be given by

$$\frac{e^{-\lambda d/v}}{1 - e^{-\lambda d/v}}.$$

This will be the ratio of those photons observed to have energy E_0 to photons observed to have the increased energy $E_0(1 + v/c)$. In principle, by varying d and measuring the change in height of the two peaks in the energy spectrum, assuming them to be separable, a determination of λ can be made. If the peaks are not separable then the centroid of the amalgamated peak will shift as d is varied. From this information, and a knowledge of the shape of the two peaks, again λ can be determined.

This method has its greatest sensitivity when

$$\frac{\lambda d}{v} \simeq 1.$$

If v is given the realistic value of 3×10^6 m s^{-1}, then, for the measurement of states with a mean life of 10^{-12} s, $d = 3 \times 10^{-6}$ m. It follows that considerable experimental difficulties are involved in applying this technique in the time range in which it is of interest.

We can best see the γ-ray energy discrimination called for by considering a particular case. Suppose ^{19}F to be bombarded with 5 MeV α-particles in order to study an excited state approximately 1·5 MeV above the ground state of ^{22}Na, produced by the reaction

$$^{19}\text{F} + {}^4\text{He} \rightarrow {}^{22}\text{Na}^* + {}^1\text{n} + Q.$$

From the mass excesses listed in Appendix A, and with the knowledge that the mass excess of ^{22}Na is $- 5 \cdot 182$ MeV, we find the Q-value of the reaction to be $- 3 \cdot 45$ MeV. Thus when the α-particle bombarding energy is 5 MeV, the ^{22}Na* recoil and the neutron share 1·55 MeV. There are two ways in which this energy can be shared, having regard to the conservation of linear momentum, when the products travel parallel to the beam. These correspond to the neutron being emitted in the forward and backward direction in the centre-of-mass system. In the laboratory system the ^{22}Na* can recoil forward with an energy of 1·18 MeV and the neutron travel backwards with an energy of 0·37 MeV. Alternatively the ^{22}Na* can recoil forwards with an energy of 0·55 MeV and the neutron also travel forward with an energy of 1·00 MeV. Considering now the former case we find that, when the recoil has an energy of 1·18 MeV its velocity v is given by

$$\frac{v}{c} = \sqrt{\left[\frac{2T}{Mc^2}\right]} = 1 \cdot 07 \times 10^{-2}.$$

Hence the Doppler shift in energy, which is equal to vE_0/c, is 16 keV when $E_0 \simeq 1 \cdot 5$ MeV. An energy shift of this magnitude can readily be measured by a lithium-drifted germanium detector.

The Doppler effect not only shifts a line but in practice considerably broadens it. There are several reasons for this. Finite target thicknesses are always involved and lead to small but variable energy losses, by the bombarding particle on entering and by the recoil on leaving. These losses cause a spread in the value of v. Again, the recoils may be projected at an angle θ to the beam direction. In that event, if the velocity v is as for the recoil projected at $0°$, then the Doppler shift is reduced and becomes

$$\frac{v}{c} (\cos \theta) E_0.$$

In fact it is likely that the velocity will be a function of θ and this leads to further broadening. Most important of all however is the fact that a finite stopping time must be involved. If this stopping time is appreciable compared to τ, then a fraction of the nuclei will de-excite in the stop when their velocity is intermediate between v and 0. For the arrangement in Figure 75 to be practicable it is therefore necessary that τ be very much greater than the stopping time of the recoil. This condition, quite apart from the smallness of d noted above, sets a limit to the technique, as in practice stopping times are approximately 5×10^{-13} s.

Although an arrangement as in Figure 75 has been used in some instances to measure mean lives down to 10^{-12} s, it is now more usual to dispense with the gap between the target and the stop, allowing the recoil to be brought to rest either in the target or in a solid backing placed in contact with the target. When the mean life of the excited state is comparable to the stopping time of the recoil in the solid material, the shape of the Doppler-shifted line depends on τ. We now proceed to discuss how an estimate of τ can be made in these circumstances.

We have to consider the recoil velocity v to vary continuously from the moment of the projection of the recoil which we take to correspond to $t = 0$. We assume that the recoil is not scattered significantly while being stopped. In an interval dt, at time t, the probability of a nucleus de-exciting will be $\lambda \, dt$ and its Doppler shift $v(t)E_0/c$. If then we are dealing with a sample of N_0 recoils, $N_0 e^{-\lambda t}$ will have survived to a time t; thus the number of photons emitted in the forward direction, and therefore having a Doppler shift $v(t)E_0/c$, will be

$$G N_0 e^{-\lambda t} \lambda \, dt,$$

where G is a geometrical factor. The average Doppler shift will therefore be

$$G \int_0^\infty N_0 e^{-\lambda t} \lambda \frac{v(t) E_0}{c} \frac{dt}{G N_0} = \frac{\lambda E_0}{c} \int_0^\infty v(t) e^{-\lambda t} \, dt.$$

This is customarily divided by the maximum Doppler shift $v_0 E_0/c$, expressed in terms of $\tau = 1/\lambda$, and written

$$F(\tau) = \frac{1}{v_0 \tau} \int_0^\infty v(t) e^{-t/\tau} \, dt.$$

To proceed further we require to know how the velocity varies with time. The stopping of a slow heavy ion in a solid material is a complicated process which has no simple theoretical description. It is found empirically that in certain cases the rate of energy loss $- dE/dx$, termed the *stopping power*, is approximately proportional to $v(t)$. When this is so we can introduce a parameter α defined by

$$-\frac{dE}{dx} = -M\frac{dv}{dt} = M\frac{v}{\alpha},$$

and write $\quad \dfrac{dv}{dt} = -\dfrac{v}{\alpha}.$

Thus $\quad v(t) = v_0 e^{-t/\alpha}$

We can now write

$$F(\tau) = \frac{1}{v_0 \tau}\int_0^\infty v_0\, e^{-t(1/\alpha + 1/\tau)}\, dt = \frac{\alpha}{\alpha + \tau} = \frac{\lambda\alpha}{1 + \lambda\alpha}. \qquad \textbf{10.6}$$

A measurement of the mean Doppler shift, and its comparison with the maximum Doppler shift, then gives the attenuation function $F(\tau)$, from which τ can be found using equation **10.6**. It is clear from the form of this equation that the method will be most sensitive when $\lambda\alpha \simeq 1$. The observed values of α for solid materials lie in the range 0·1–1 ps and hence the DSAM technique is most suited for measurements of half-lives in the neighbourhood of a picosecond.

10.13   **The measurements of level widths**

We noted in section 10.7 that the product of the mean life of the level and its width equalled $\hbar$. Thus the measurement of the level width provides essentially the same information as the measurement of the mean life. However level-width measurements can sometimes be achieved for states where the mean life is too short to be measured by the methods discussed in section 10.12.

We begin by considering quantitatively the nature of the problem of measuring the level width in a particular case. For a nucleus with A approximately 125 and a level at say 200 keV, the 'single-particle' estimates of width quoted in section 10.7 become

$\Gamma E1 = 1{\cdot}38 \times 10^{-2} \text{ eV},$

$\Gamma E2 = 0{\cdot}98 \times 10^{-8} \text{ eV},$

$\Gamma E3 = 4{\cdot}6 \times 10^{-15} \text{ eV},$

$\Gamma M1 = 1{\cdot}68 \times 10^{-4} \text{ eV},$

$\Gamma M2 = 1{\cdot}2 \times 10^{-10} \text{ eV},$

$\Gamma M3 = 5{\cdot}44 \times 10^{-17} \text{ eV}.$

These estimates depend of course on the nuclear model from which the wave functions have been taken to calculate the matrix element, and should be considered only as a guide to likely orders of magnitudes of level widths. It is clear that it is not practical to explore the resonances by a direct method, either for example by measuring with sufficient resolution the emitted photon spectrum or by measuring the absorption by the nucleus of a beam of photons of sufficiently limited energy spread. The resolution of available γ-ray spectrometers, as was seen in section 10.12, is many orders of magnitude too low. Photon beams in the energy range of interest are not available with anything approaching the degree of energy homogeneity that would be necessary for a direct absorption experiment. It might seem that the photon emitted from a transition to the ground state would be strongly absorbed by an unexcited nucleus of the same isotope and that this might give useful information about level widths. Such a process would correspond to *resonance fluorescence* in atomic physics, which is an easily observed and well-studied phenomenon. In the nuclear context however, the study of resonance fluorescence presents severe experimental difficulties. These arise because the photon emitted by an excited nucleus does not in fact take away the full transition energy. To conserve linear momentum the emitting nucleus must recoil with a small but finite kinetic energy. Thus, the photon momentum being E_γ/c, we have

$$\sqrt{(2M_R T_R)} = \frac{E_\gamma}{c},$$

and therefore the kinetic energy T_R of recoil of the nucleus, mass M_R, is given by

$$T_R = \frac{E_\gamma^2}{2M_R c^2}.$$

Therefore $\quad E_0 = E_\gamma + \dfrac{E_\gamma^2}{2M_R c^2}.$ \hfill 10.7

The same consideration arises in the absorption process. The absorbing nucleus must recoil to take up the linear momentum of the incident photon, and hence the whole photon energy is not available as excitation energy. For absorption

$$E_\gamma = E_0 + \frac{E_\gamma^2}{2M_R c^2}.$$

Taking for illustration the case of A approximately 200, E_0 approximately 400 keV, we find

$$\frac{E_\gamma^2}{2M_R c^2} \simeq 0{\cdot}9 \text{ eV}.$$

This shift in energy which occurs both in the case of emission and absorption is seen to be much greater than even the greatest estimated natural width, that of E1, quoted above. However, the emitting or absorbing nucleus is incorporated in an atom which has thermal motion. The thermal velocity gives rise to a Doppler shift which broadens the line. The full Doppler width is given by

$$2\Delta = 2E_\gamma \sqrt{\left[\frac{2kT}{M_R c^2}\right]}.$$

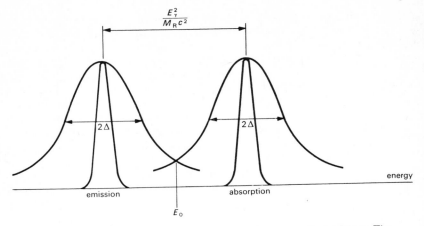

Figure 76 Shift of emission and absorption lines due to recoil of nucleus. The intrinsic width of the line (on a reduced vertical scale) is shown in the middle of the peaks, which are broadened by the Doppler shift arising from thermal atomic motion.

In our present illustration this means $2\Delta = 0.4$ eV. The relationship of the emission to the absorption energy spectrum is shown in Figure 76. It is clear that the overlap, which will determine the extent to which resonance fluorescence takes place, is extremely small.

There are several possible methods for restoring the recoil energy and increasing the overlap. For example, if the source is moved relative to the absorber with velocity v, the emitted photon energy will be Doppler shifted by an amount $(v/c)E_\gamma$. If we calculate the velocity necessary to bring the emission and absorption spectra in Figure 76 into coincidence we find $v = 5.1 \times 10^2$ m s^{-1}. Moon (1951) succeeded in achieving speeds of this magnitude by attaching a source to the tip of a high-speed rotor and demonstrated resonance fluorescence for 418 keV photons emitted by ^{198}Hg.

A second possibility for restoring the recoil energy arises when the photon is emitted so soon after a β- or γ-emission that the nucleus is still recoiling from the effect of the previous emission. For the values of A and E_0 considered above, the nucleus could have sufficient recoil momentum following the emission of a γ-ray of at least 400 keV or a β-particle of energy in excess of 140 keV. However, even when a source with a suitably fast transition can be found, the fact that the β-energy spectrum is continuous and also the requirement that the recoil has to be travelling at the correct angle with respect to the photon direction very severely limits the number of recoils which would contribute to fluorescent absorption.

The possibilities of experimentally studying nuclear resonance fluorescence were however completely altered by the discovery in 1958 by Mossbauer of 'recoilless' emission of γ-rays. This effect, now known as the *Mossbauer effect*, relies on the emitting and absorbing nuclei being incorporated in a crystal lattice. On the view that the lattice constitutes the quantum system, there is a certain minimum energy, namely that of one quantum, by which the lattice internal energy may be increased. If the kinetic energy of recoil of the nucleus following γ-emission is less than this minimum then there is no relative motion of the nucleus with respect to its neighbours possible and the whole lattice must recoil. On substitution of the lattice mass for the nuclear mass in equation **10.7** we see that the difference between E_γ and E_0 is now immeasurably small. The same consideration holds for absorption. We thus can have, with the source and the absorber in a suitable physical form, a complete overlap of the emission and absorption curves of Figure 76. When this coincidence has been achieved, a direct measure of the widths of these resonant curves becomes possible by introducing a relative motion between source and absorber. This motion leads to a Doppler shift and consequent separation of the emission and absorption curves. The result of such an experiment is shown in Figure 77. It is seen that in the case of the 129 keV line of ^{191}Ir a velocity of 100 mm s^{-1}, which corresponds to a Doppler shift of approximately 50×10^{-6} eV, is sufficient to make the overlap of the emission and absorption curves negligibly small. This transition in ^{191}Ir, which was that used by Mossbauer in establishing the effect, has been shown by a delayed coincidence experiment to have a half-life of $1\cdot3 \times 10^{-10}$ s. It is seen that the

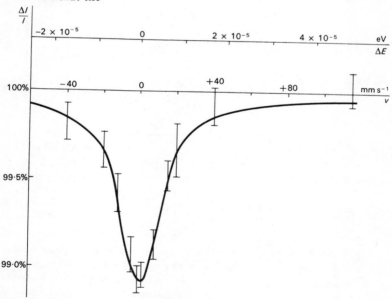

Figure 77 The results of the original Mossbauer experiment. The transmission of 129 keV γ-radiation from ^{191}Ir through an iridium absorber is measured as a function of relative velocity v of source and absorber

estimate of a width of about 5×10^{-6} eV which one would make using the uncertainty principle is in agreement with the measured transmission through the sample in Figure 77. The experimental curve which arises from moving an emission line of width Γ over an absorption line of the same width should have Lorentzian shape and line width 2Γ.

The special conditions which have to be satisfied by a transition for the Mossbauer effect to occur (namely that the transition energy should be small, so

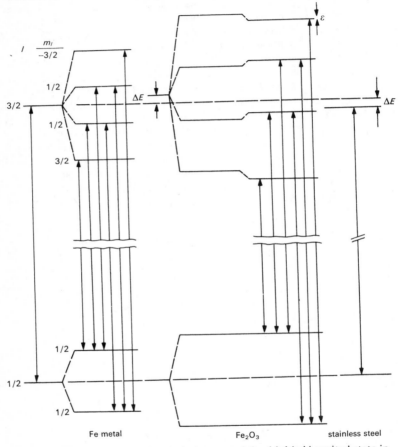

Figure 78 The splitting of the ground state and the 14·4 keV excited state in ^{57}Fe. In Fe metal there is a large internal magnetic field with which the nuclear dipole moment interacts, but because of the cubic symmetry of the lattice there is no electric field and hence no electric quadrupole interaction. In Fe_2O_3 there is an electric field gradient whose interaction with the electric quadrupole moment of the nucleus affects the magnetic substate of the upper level. In stainless steel the time-averaged magnetic field is zero and there is no electric field gradient. ΔE represents small shifts due to chemical binding effects

that the recoiling nucleus is locked in the lattice, and that the level width should be such that the Doppler-shift relative velocities are of a practical magnitude) will mean that the method will always be limited to very few transitions. About sixty suitable transitions have been found to date. In each case the level widths are as would be predicted for the measured half-lives.

While the Mossbauer effect has not added significantly to what was already known via the lifetime measurements, it has made possible hyperfine-structure measurements of very great sensitivity which lead to accurate determinations of nuclear moments. In this connection the nuclide ^{57}Fe has received much attention. The 14·4 keV magnetic dipole transition in ^{57}Fe has a half-life of $1·4 \times 10^{-7}$ s and a width of $4·6 \times 10^{-9}$ eV. In this case, the relative velocity of source and detector as small as 1 mm s^{-1} produces a Doppler shift large enough to affect the fluorescent absorption. When the Fe nucleus is embedded in stainless steel, the time-averaged magnetic field at the nucleus is zero. The levels are therefore not split and the emitted radiation is strictly monochromatic. When now an absorber of Fe metal is used, the large internal magnetic field splits the levels as shown in Figure 78. The metallic crystal lattice is such that there is no electric field at the Fe nucleus. In Fe_2O_3, however, there is an expected electric field gradient which causes shifts in the magnetic substates large enough to measure by the Mossbauer technique. The measured absorption curve in Figure 79 shows the accuracy with which the level spacings may be measured and values of μ and Q determined, as discussed in Chapter 8.

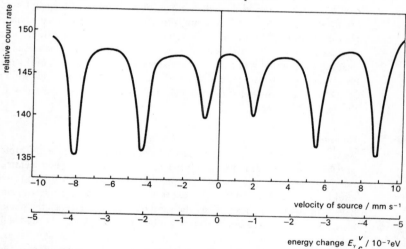

Figure 79 Absorption in Fe_2O_3 of the 14·4 keV γ-rays emitted in the decay of $^{57}Fe^*$ formed by the decay of ^{57}Co incorporated in a stainless steel source. The six dips correspond to the six transitions of Figure 78

10.14 Transition strengths

The probability or strength of a transition, we have now seen, can be expressed in terms of a decay constant λ, a width Γ or a mean life τ. It is very often, however, convenient to take the single-particle predictions of section 10.6 as standards. When the observed width of a transition is divided by the appropriate single-particle width the ratio is said to give the transition strength in Weisskopf units.

When the measured transition strengths are examined it is found that, except for E2 transitions, the strengths in Weisskopf units are less than unity. In the case of E1 transitions and light nuclei the strengths may be as small as a few per cent. For E2 transitions, on the other hand, both in light and heavy nuclei there are many examples of strengths from ten to a hundred times the single-particle value.

An explanation of the enhancement of E2 transitions is to be sought in terms of the collective model of section 9.3. When this model is used instead of the single-particle model to calculate the transition matrix elements it is found that

$$B(\text{E2}) = \frac{15}{32\pi} e^2 Q_0^2 \frac{(I+1)(I+2)}{(2I+3)(2I+5)}, \qquad \textbf{10.8}$$

for a transition from a state spin $I + 2$ to a state spin I. In this formula Q_0 is the static quadrupole moment. A large transition strength is therefore evidence of the existence of a large quadrupole moment.

The same reduced transition probabilities are involved in the cross-section for Coulomb excitation and the enhancement of E2 transitions is again found. From the measured values of $B(\text{E2})$, using formula **10.8**, the static quadrupole moment Q_0 can be found. This provides a very important alternative method to that discussed in Chapter 8. Whereas in the earlier method the interaction of the quadrupole moment was with an internal electric field gradient, in the method now under consideration the interaction is with the radiation field. More reliance can be placed on the radiation-field calculation than on the calculation of internal electric field gradients.

10.15 Unbound states

We have, in discussing radiation widths, been concerned with bound states; the possible modes of decay we therefore had to consider were gamma emission, internal conversion and internal pair production. The total width of the bound level is the sum of the widths of these three processes, whose relative contributions have been discussed above. When the excitation energy exceeds the nucleon separation energy, and nucleon emission becomes possible, then the total width contains contributions from Γ_n, Γ_p, Γ_α, etc., the widths to be associated with particle emission. Since the lifetime of the state against decay by particle emission will be of the order of v/R, v being the particle velocity in the nucleus, R the nuclear radius, we find the mean life τ for this mode of decay to be about 10^{-21} s and the level width therefore to be hundreds of thousands of electronvolts. Thus the total width is dominated by the particle width, the gamma width making a negligible contribution.

The assumption that the specifically nuclear force between two nucleons is the same for nn, pp, np pairs results in important simplifications in nuclear theory. This hypothesis is referred to as the *charge independence* of nuclear force. If this is a valid hypothesis, it is to be expected that there will be similarities in the level structure of assemblies of a given number of nucleons, since the exchange of a proton for a neutron or vice versa should not affect the binding energy. We find in fact similarity in the level structure of mirror nuclei such as ^{7}Li and ^{7}Be. This can be seen in Figure 80. If we allow for the difference in Coulomb energy and the difference between neutron and proton rest masses, neither of which effects arises from nuclear forces, then the agreement in level position on an absolute scale is even better.

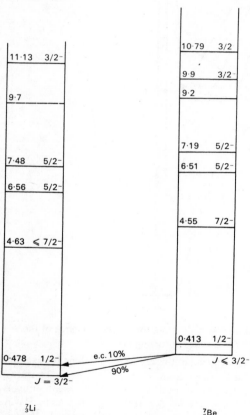

Figure 80 Energy level diagrams of ^{7}Li and ^{7}Be showing similarities in structure

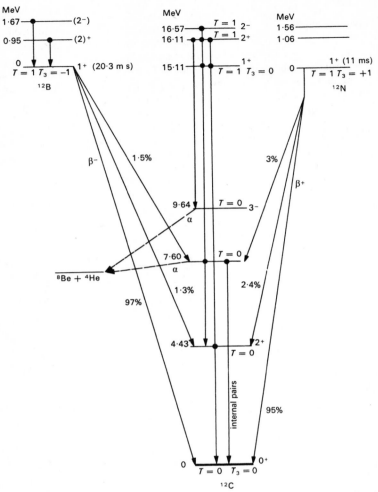

Figure 81 Isobaric triad with $A = 12$. Note the correspondence of the $T = 1$ states and the additional $T = 0$ states occurring only in ^{12}C

When we consider even-A nuclei in this connection, taking for example the isobars with $A = 12$ whose levels are shown in Figure 81, we find a state in the central nucleus corresponding to the ground state of both neighbouring isobars. This state is said to be the *isobaric analogue* of these ground states. Having regard to the Pauli exclusion principle, we see that the ground states of ^{12}B and ^{12}N must have the nucleons in spatial states with spins oriented as shown schematically in Figure 82. The analogue state of ^{12}C is shown for comparison. It is clear however

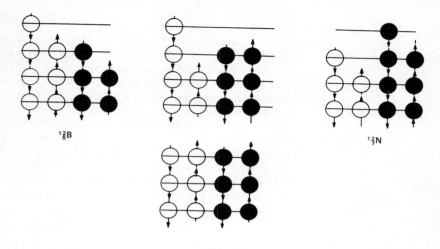

$^{12}_{5}B$

$^{12}_{7}N$

$^{12}_{6}C$

Figure 82 Schematic representation of $T = 1$ states in three members of an isobaric triad and $T = 0$ state in one member only. The Pauli exclusion principle dictates the number and spin orientation of the particles which may occupy each level

that a more symmetrical and hence more tightly bound arrangement is possible in ^{12}C but that this arrangement is not available to the other isobars because of the operation of the exclusion principle.

An elegant description of this situation can be given in terms of *isospin* (or *isobaric spin*), a concept first introduced by Heisenberg (1932) when it was misleadingly referred to as isotopic spin. Neutrons and protons are treated as two states of a nucleon, the charge of the proton and the mass difference between the proton and the neutron, believed to be electromagnetic in origin, being ignored. A new quantum vector **T**, the isospin, is introduced and given the property, similar to that of ordinary spin, of having allowed values $+\frac{1}{2}$ and $-\frac{1}{2}$ for its components T_3, along a specified direction. $T_3 = +\frac{1}{2}$ we associate with a proton and $T_3 = -\frac{1}{2}$ with a neutron (this, the currently accepted convention, is the opposite to that originally chosen).

Let us now take a pair of nucleons; the isospin of the pair is given by $T = 1$ or $T = 0$ from the addition of the two isospin vectors, carried out according to the rules applied to ordinary spin. There are then three possible components T_3 to be associated with $T = 1$ namely $T_3 = +1, 0, -1$. To $T_3 = +1$ will correspond two protons with antiparallel spin, to $T_3 = 0$ a neutron–proton pair, again with antiparallel spin and to $T_3 = -1$ a pair of neutrons with antiparallel spin. $T = 0$ will be a singlet, having only the possibility of $T_3 = 0$, which will correspond to a neutron–proton pair with parallel spin. On this interpretation the ground state of the deuteron with ordinary spin 1 has thus $T_3 = 0, T = 0$.

In the general case each proton in a nucleus will introduce $T_3 = +\frac{1}{2}$, each neutron will add $T_3 = -\frac{1}{2}$ and hence for the nucleus $T_3 = \frac{1}{2}(Z - N)$. For example, for ^{12}B $T_3 = -1$, for ^{12}C $T_3 = 0$, for ^{12}N $T_3 = +1$. It is now seen that the low-lying states in ^{12}C which have no analogues in the isobars are states with $T = 0$. The analogue state at 15·1 MeV in ^{12}C is the lowest-lying $T = 1$ state in that nucleus. At higher energies in all three isobars there will be $T = 2$ states, the lowest-lying of which will be expected to be analogues of the ground state of ^{12}Be and ^{12}O.

The isospin substates will be degenerate only in so far as the Coulomb interaction is neglected. When the Coulomb interaction is taken into account the level splits in the same way as magnetic substates separate in a magnetic field. When the splitting becomes appreciable, as would be expected in nuclei of high atomic number, then T ceases to be meaningful. In view of this expectation it has come as a surprise that isobaric analogue states can still be recognized in nuclei with A-values as high as ninety.

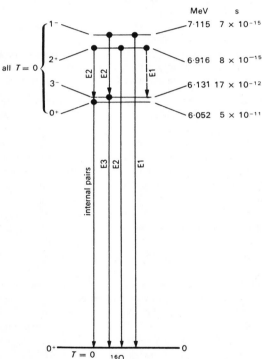

Figure 83 Transitions between some levels of ^{16}O to illustrate the operation of the isobaric spin selection rule in electric dipole transitions. The E1 transition from the 6·916 MeV level to the 6·131 level is not observed and the E1 transition from the 7·115 MeV level to the ground state has a half-life several orders of magnitude longer than the Weisskopf estimate, as can be seen from Figure 69

It is found that, in radiative transitions, alongside the selection rules of section 10.5.3 we have to place isospin selection rules. In a radiative transition in which only one nucleon changes its state $\Delta T = 0$ or ± 1. This does not lead to any immediately practical result as $T = 2$ states lie very high in the level diagrams of light nuclei. However, for electric dipole transitions in self-conjugate nuclei (i.e. having $Z = N$) there is the special rule that $\Delta T = \pm 1$. This is of practical importance as very many low-lying states have $T = 0$, and $T = 0 \rightarrow T = 0$ is forbidden by this rule. This forbiddenness arises because in both initial and final states the neutrons and protons are occupying identical nucleon states. There is thus in the transition no separation of neutron and proton distributions, which is necessary for there to be an induced dipole moment. The operation of this isobaric spin selection rule can be seen in Figure 83 in the case of ^{16}O. It is seen that the rule is not inviolable but that if it is not obeyed the transitions are considerably inhibited.

0.17 Summary

We have seen that the broad picture now held of nuclear excited states is of low-lying collective states, vibrational in nature in the case of the spherical nuclei, rotational in the case of the deformed nuclei, with single-particle states lying above. Still higher lie states corresponding to excitations of the nuclear core as distinct from collective effects in the outer shell. These higher states are involved in photonuclear reactions, and, lying above the particle separation energy, they are capable of decaying by particle emission.

In light nuclei similarities are found in the level structures of isobars which encourages the introduction of a further quantum number, isobaric spin, to join spin and parity in the designation of a nuclear level.

Chapter 11
Nuclear Fission

11.1 Introduction

In section 5.11 a brief and formal discussion was given of the process of fission, in which a heavy nucleus breaks into two main fragments with the release of an amount of energy which even by nuclear standards is spectacular. In sufficiently heavy nuclei fission takes place spontaneously from the ground state, competing as a decay process with α-decay. A theoretical treatment can be given in terms of the concepts introduced in α-decay theory. Fission may also be induced by bombarding a heavy element with neutrons or other nuclear projectiles, or by exposing the element to a beam of high-energy X-rays. *Induced fission* is a special case of a nuclear reaction, in which a compound nucleus is formed in an excited state and fission is one of the several possible modes of decay. All of these decay modes, including fission, will have widths as defined in section 10.7. A discussion of induced fission more properly finds its place in the companion volume, W. M. Gibson *Nuclear Reactions*, Penguin, 1971. In the present chapter we are concerned with fission as a nuclear decay process.

11.2 Experimental aspects of fission

Experimentally, fission is observed to involve the emission of two massive fragments, each having kinetic energy of about 75 MeV. In the case of spontaneou fission the fragments are not usually of equal mass; the fission is then said to be *asymmetric*. In the case of induced fission the asymmetry in mass decreases with increase in bombarding energy and the fission may become *symmetric*. In Figure 8 are plotted the mass distributions of fission fragments for the isotopes ^{234}U, ^{236}U and ^{240}Pu, the compound nuclei formed by neutron capture into ^{233}U, ^{235}U and ^{239}Pu. Note that the ratio of the yields at the asymmetric peaks to the symmetric yield is so large that a logarithmic scale has been used in the presentation.

All of the nucleons of the original nucleus are not usually accounted for by the composition of the two fragments. Because of the increase in neutron excess with mass number of stable nuclei, there are more neutrons available than the fission fragments can stably contain. As a consequence, not only are the fission fragment created with Z- and N-numbers lying well off the stability area on the nuclear chart, by virtue of excess neutrons, but in addition free neutrons are released 'promptly' in the process. The energy spectrum of these neutrons peaks at 1 MeV and falls off exponentially, the 'tail' extending to about 15 MeV. The emission of

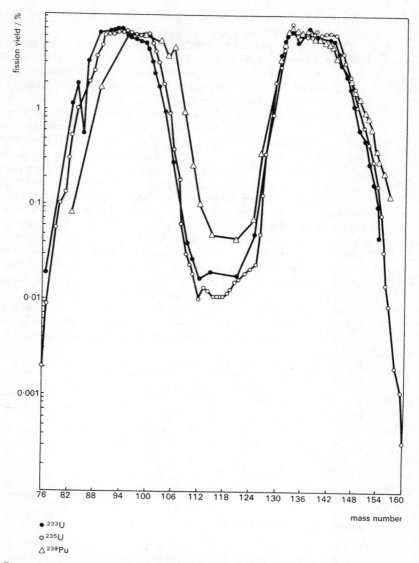

Figure 84 The mass distribution of fission fragments in the slow-neutron fission of ^{233}U, ^{235}U and ^{239}Pu

prompt neutrons as products of the fission events and their potentiality for inducing further fission events make a chain reaction a practical possibility.

When only two charged fragments are emitted the fission is termed *binary*. In a relatively small percentage of the fission decays, usually about a few tenths of one per cent, a third charged particle such as an α-particle may be emitted; the process is then termed *ternary fission*.

The emission of γ-rays is a further accompaniment of the fission process. This is to be expected as a consequence of the rapid charge rearrangement which takes place as fission occurs. To sum up, we note that the *primary* process consists of the emission of the fission fragments together with neutrons and prompt γ-rays and very occasionally other light particles. There are *secondary* effects, also involving the release of considerable amounts of energy, which occur later. An important secondary effect is the β-decay of the fission fragments. As can be seen from Figure 84, very many β-active nuclides are formed. The half-lives range from a fraction of a second (e.g. ^{85}As: 0·43 seconds) to millions of years (e.g. ^{129}I: 1·6 x 10^7 years). Many of the fragments when formed are so far from stability that several successive decays, each involving a neutron-to-proton switch, and hence the emission of a β$^-$ particle, are necessary before stability is achieved. The nuclei involved in these decay chains may of course be formed in excited states and γ-rays may therefore be involved in the decay processes. Consequently it is found that β-rays and γ-rays continue to be emitted at a decreasing rate from fission fragments removed from fissile materials. We noted in section 6.1 that neutrons are emitted from some nuclides following the β-decay of fission fragments. These constitute secondary *delayed* neutrons and play an important role in the control mechanisms of nuclear reactors.

The distribution of the fission energy among the primary and secondary particles and radiation is of great importance in the practical application of nuclear fission in reactors and weapons and is given for interest in Table 7.

Table 7 Distribution of Fission Energy

	MeV
Kinetic energy of fission fragments	165 ± 5
Instantaneous γ-ray energy	7 ± 1
Kinetic energy of fission neutrons	5 ± 0·5
β-particles from fission products	7 ± 1
γ-rays from fission products	6 ± 1
Neutrinos from fission products	10
	200 ± 6

11.3 Elementary theory of fission

In Figure 85 we have plotted the potential energy of the heavy nucleus as a function of the distance of separation between centres of the fission fragments, here regarded as being of spherical shape with radii R_1 and R_2. From very large separation distances down to a distance $R_1 + R_2$, assuming the spheres to remain undistorted until they touch, this curve is a hyperbola since

$$V = \frac{e^2 Z_1 Z_2}{r}.$$

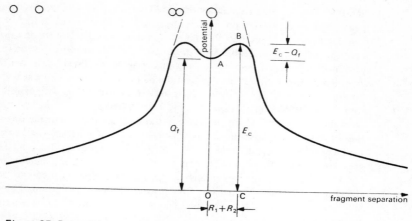

Figure 85 Potential energy against fission fragment separation

From the separation at which the fragments touch down to zero separation, where they have completely merged to form one nucleus, there is no simple way of finding the form of this potential curve. We do know however that it must pass through A where $OA = Q_f$ is given by the difference between the nuclear mass and the sum of the masses of the fragments.

To establish the relative lengths of $OA = Q_f$ and $CB = E_c$ let us consider the particular case of

$$^{236}_{92}U^{144} \rightarrow {}^{95}_{36}Kr^{59} + {}^{141}_{56}Ba^{85}.$$

The excess masses of ^{236}U, ^{95}Kr and ^{141}Ba being 42·52 MeV, $-$ 58·34 MeV and $-$ 79·97 MeV respectively, we find that $Q_f = 180\cdot83$ MeV. If we take the nuclear unit radius R_0 (see section 3.14) to be 1·48 fm, in order to evaluate $R_1 + R_2$, and extrapolate the hyperbola, as shown in Figure 85, to the point of contact of the spheres, the maximum of the curve is found to be

$$\frac{e^2 Z_1 Z_2}{R_0(A_1^{\frac{1}{3}} + A_2^{\frac{1}{3}})} = 201 \text{ MeV.}$$

It is however to be expected that the deformation of the fragments and the details of the fall off of nuclear force will remove the sharpness and round the curve off to a lower maximum value. We can invoke a very important difference in the behaviour of the two isotopes ^{235}U and ^{238}U to establish the difference $E_c - Q_f$. When natural uranium (99·27 per cent ^{238}U; 0·72 per cent ^{235}U) is bombarded with slow neutrons, fission is induced in the ^{235}U isotope only; when, however, the neutron energy is increased to 1·2 MeV, fission is induced in ^{238}U. We interpret this as establishing that there is a threshold energy for the process

$$^{238}U + n \rightarrow {}^{239}U^* \rightarrow X + Y,$$

and that there is no corresponding threshold when the lighter isotope captures a neutron. If we now, from the mass values, calculate the extent to which the compound nuclei ^{236}U and ^{239}U are excited following the capture of a slow neutron, we find that it is 6·5 MeV in the case of ^{236}U and only 4·8 MeV in the case of ^{239}U. This difference in excitation energy is explained by the neutron completing a neutron pair in ^{236}U and the pairing energy becoming available for internal excitation. The neutron added to ^{238}U does not have the possibility of forming a pair. Now Q_f in the case of ^{239}U is 181·14 MeV, very little different from the value for ^{236}U. The energy curve at large separations of the fragments should be the same for both isotopes. The existence of a threshold in the case of the heavier isotope, and its absence in the case of the lighter, is then satisfactorily explained if $E_c - Q_f$ is approximately 6 MeV. The addition of a neutron to the lighter isotope then provides enough energy for the fission fragments to 'go over the potential barrier'. In the case of the heavier isotope the addition of a neutron still leaves the excitation energy such that a barrier is presented to the outgoing fission fragments.

The threshold for induced fission mentioned above is only an apparent threshold. Barrier penetration, while unlikely, still has a finite probability, so that the onset of fission will not be expected to commence abruptly at a well-defined energy but to increase continuously with increasing excitation energy as the width of the barrier to be penetrated decreases.

In the case of nuclei lighter than uranium, $E_c - Q_f$ will be greater and fission will be more difficult to induce. With heavier nuclei on the other hand, $E_c - Q_f$ will be expected to decrease; when it attains small values spontaneous fission by barrier penetration will become more likely, the half-life for the process getting progressively shorter as $E_c - Q_f$ diminishes.

A deeper understanding of the phenomenon of fission requires a study of nuclear behaviour in the neighbourhood of the potential-energy maximum. This can only be undertaken in terms of a specific nuclear model. N. Bohr and Wheeler (1939) laid the foundations for tackling the problem in terms of the liquid-drop model. The nucleus in its ground state was likened to a spherical drop of incompressible liquid of uniform density and uniformly distributed charge. When deformed from its spherical shape, its potential energy was assumed to be determined by the surface- and Coulomb-energy terms defined in section 5.5. We now proceed firstly to investigate in terms of these assumptions the effect of a small deformation in order to explore the stability of the spherical shape, and secondly to consider the effect of deformations large enough to take the system through the potential-energy maximum.

11.4   Nuclear stability: small deformations

The stability of the electrically charged drop against small disturbances will be determined by the size of the increase in surface energy, compared to the decrease in Coulomb energy, caused by the departure from spherical shape. These energy changes can be calculated in purely classical terms. Assume the small deformation to be from a spherical shape, radius R, to the shape of a spheroid with equal semi-axes b and the third semi-axis equal to c. The incompressibility of the liquid demands constancy of volume; hence we have

$$\tfrac{4}{3}\pi R^3 = \tfrac{4}{3}\pi c b^2.$$

We now introduce ϵ', writing

$$c = R(1 + \epsilon'), \qquad b = \frac{R}{(1+\epsilon')^{\frac{1}{2}}},$$

where ϵ' is a small quantity. On deformation, the surface energy, due to the increase in surface area, is increased from

$$a_s A^{\frac{2}{3}},$$

the value we attributed in section 5.5 to the spherical nucleus, to

$$a_s A^{\frac{2}{3}}(1 + \tfrac{2}{5}\epsilon'^2),$$

higher powers of ϵ' being neglected. The Coulomb energy, on the other hand, is reduced from the spherical value of

$$a_C \frac{Z^2}{A^{\frac{1}{3}}}$$

to

$$a_C \frac{Z^2}{A^{\frac{1}{3}}}\left[1 - \frac{1}{5}\epsilon'^2\right],$$

to the same order of accuracy. We therefore conclude that, for the particular form of deformation here being considered, the drop will be stable or unstable when slightly disturbed depending on whether $\tfrac{2}{5}a_s A^{\frac{2}{3}}$ is greater or less than $\tfrac{1}{5}a_C Z^2/A^{\frac{1}{3}}$, i.e. whether Z^2/A is less or greater than $2a_s/a_C$. This critical value we denote by

$$\left[\frac{Z^2}{A}\right]_c.$$

For the values of a_s and a_C introduced in section 5.5 it is approximately equal to 44. We recollect that mass–energy considerations alone in section 5.11 lead to the conclusion that symmetric spontaneous fission 'over the barrier' became possible for

$$\frac{Z^2}{A} \gtrsim 50.$$

Asymmetric fission may be possible for lower values of this quantity. Also, allowing for the probability for barrier penetration, we will not expect a sudden onset of spontaneous fission at a particular value of Z^2/A. However, it is very convenient to form the ratio of Z^2/A to $(Z^2/A)_c$ for any particular nucleus and to regard this ratio as a measure of the stability of the nucleus against spontaneous fission. It is usually denoted by x and called the *fissionability parameter*. In Figure 86 are plotted some measured values of spontaneous-fission half-lives against Z^2/A to show the essential validity of these concepts.

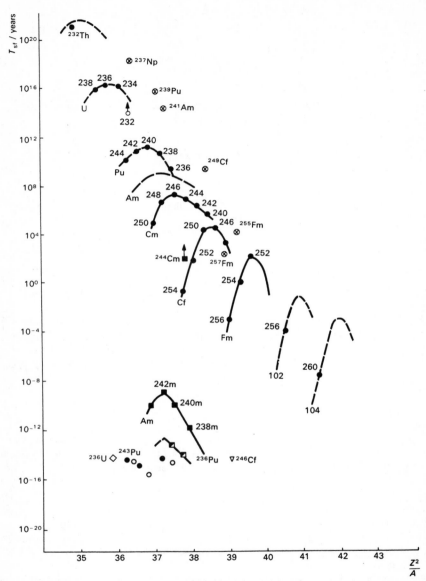

Figure 86 Plot of half-life for spontaneous fission against Z^2/A for heavy nuclides. Some shape isomers (see section 11.8) are included

11.5   Nuclear stability: large deformations

We now consider a nucleus whose fissionability parameter x is less than unity and which therefore has metastability against spontaneous fission. We wish to consider circumstances in which it suffers a deformation large enough for the long-range Coulomb forces to dominate the short-range nuclear forces which give rise to the surface energy. The nucleus will then proceed to undergo fission. The classical treatment of large deformations of a charged liquid drop is not in principle difficult but the mathematical complexities are very formidable. We can follow the procedure outlined in section 9.5.1 in connection with nuclear vibrations and express the deformed surface as a combination of spherical harmonics. The radius vector R will then be given by

$$R(\theta, \phi) = R\left[1 + \sum_{\lambda=1}^{\infty} \sum_{\mu=-\lambda}^{+\lambda} \alpha_{\lambda\mu} Y_{\lambda\mu}(\theta, \phi)\right].$$

In the face of the complications involved for completely general deformations, we limit consideration to the particular deformations in which $\mu = 0$ (which makes the radius vector independent of ϕ, and hence the surface is always that of a solid of revolution) and we take $\lambda = 2$ and $\lambda = 4$ as the only harmonics. These values of λ, as do all even values, lead to a shape which is symmetrical with respect to the xOy plane. This can be seen from Figure 87, in which are illustrated the cases $\lambda = 2, 3, 4$. We thus have for our deformed shape

$$R(\theta) = R[1 + \alpha_2 P_2(\cos \theta) + \alpha_4 P_4(\cos \theta)],$$

where $P_2(\cos \phi)$ and $P_4(\cos \phi)$ are as quoted in section 6.4. The next step is then to calculate the potential energy corresponding to different values of the two parameters α_2 and α_4 which are now being used to define the deformed shape. When this has been carried through, the result can be presented three-dimensionally, the energy plotted vertically against α_2 and α_4 plotted in the horizontal plane. To convey the situation two-dimensionally, we can plot contours as on a geographical map. The general form of the result can be seen in Figure 88. The spherical nucleus sits, as it were, in a bowl on an elevated plateau ringed around by mountains. There is a pass or saddle through these mountains to a lower plain lying beyond. This saddle represents the easiest (in the sense that least deformation energy is required) route from the original metastability of the nucleus to fission. The highest point in the pass is called the *saddle point*. It is found that at the saddle point the nucleus has not yet reached the point of separation into two parts although a waist has developed. The later stage when the separation is reached, is termed *scission point*.

More elaborate calculations introducing values of λ as large as $\lambda = 18$ have been carried through and the potential mapped out with higher accuracy. The form of the results remains the same. We therefore now have the picture of the

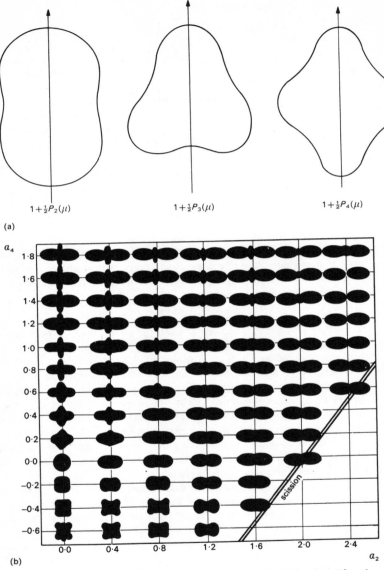

$$1 + \tfrac{1}{2}P_2(\mu) \qquad\qquad 1 + \tfrac{1}{2}P_3(\mu) \qquad\qquad 1 + \tfrac{1}{2}P_4(\mu)$$

(a)

(b)

Figure 87(a) Deformed spherical shapes comprising $\lambda = 2$, $\lambda = 3$ and $\lambda = 4$ deformations. Bodies are solids of revolution about the axes indicated. (b) Shapes of nuclear surfaces comprised of $\lambda = 2$ and $\lambda = 4$ deformations only. The figures possess rotational symmetry about the horizontal axis. The scission line is given by $\alpha_4 = -2.5 + 1.25\alpha_2$

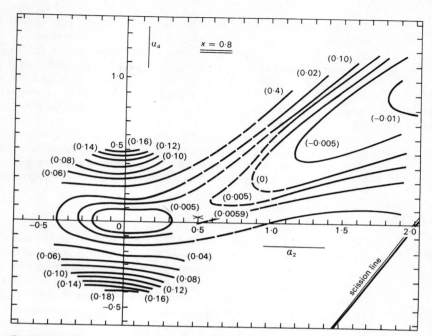

Figure 88 Contour map of potential energy of surface as a function of α_2 and α_4, all other αs being zero. Deformation energy is measured relative to spherical shape

nucleus being raised to an excited level by, let us say, neutron capture. This level will be in a region of high level density about 8 MeV above the ground state. The internal energy then has to be reorganized to deformation energy so that the nucleus takes up the shape appropriate to the saddle point. Thereafter it can pass back towards its original state or carry over towards scission.

11.6   Transition states

A. Bohr (1955) developed the idea that, the internal energy of the nucleus having been largely converted to energy of deformation and assuming that fission is taking place not too far above threshold, the nucleus will be 'cooled' at the saddle point. As a consequence its level structure will resemble that close to the ground state of a normal nucleus. Since it is highly deformed it should be expected to show collective rotational levels. In proceeding from the ground state to the scission point, the nucleus will have to go through the 'ground state' or one of the comparatively small number of collective levels at the saddle-point, levels which are said to be associated with *transition states*. The number of available transition states will control the rate at which fission proceeds and the spins associated with these states will impose angular-momentum conditions on the system. The concept of transition states has proved very fertile and there is good evidence for their existence. For example, the fission cross-section as a function of excitation energy of the fissile nucleus shows a series of plateaux which can be interpreted as corresponding to a succession of new transition states entering the process. Further, a value of K, the spin about the symmetry axis of the deformed nucleus (see section 9.4), can be investigated by studying the angular distribution of the fission fragments (which will be emitted along the symmetry axis) when the spin of the system is aligned in a nuclear alignment experiment or can be determined from the kinematics in the case of fission induced by a beam of energetic particles.

11.7 Scission point

No convincing theoretical description is yet available of the scission process, i.e. the final tearing of nuclear matter as the fragments separate. The details of the tearing will decide whether the fission is binary, ternary or involves more than three products. A considerable amount of experimental evidence for ternary fission has now accumulated. If we take the extreme (but not customary) view of regarding the neutron as a fission fragment, then two fission fragments plus a neutron strictly would constitute 'ternary fission'. A study of the angular distribution of prompt neutrons with respect to the direction of the main fission fragments reveals that a proportion of the neutrons are emitted isotropically. These are the ternary neutrons. The non-isotropic neutrons will be those emitted, following scission, from the recoiling fragments. Light charged fragments have also been observed to be emitted at scission. We noted in section 11.2 that of a thousand fission events about two or three involved the emission of an α-particle in addition to the main fragments. In Table 8 the relative frequency of occurrence of fission accompanied by the emission of other charged fragments is shown. The figures relate to long-range particles emitted by ^{252}Cf decaying by spontaneous fission.

Evidence for 'true' ternary fission involving the emission of three equal fragments has been sought by looking for coincidences in counters set to detect three coplanar fragments emitted in directions 120° apart. The results of this difficult experiment are claimed to establish that the rate of 'true' ternary fission to binary fission is about 10^{-6} in ^{240}Pu* and ^{242}Pu*. This experiment also indicated that the fission fragment masses lay in the ranges 30–40 a.m.u. and 50–60 a.m.u. However, careful radiochemical examination has not confirmed the presence of the predicted amounts of material with these mass values in the fission fragments. The occurrence of 'true' ternary fission, which is a matter of great importance for fission theory, is therefore in our present state of knowledge an open question.

Finally, we note that there is no satisfactory theoretical explanation available of the asymmetric distribution of mass between the fission fragments. If the deformation could be treated in full generality and not subject to the restrictions

Table 8 Long-Range Particles Emitted in Spontaneous Fission of ^{252}Cf

Particle	Measured relative intensity	Most probable energy/MeV	High-energy cut-off/MeV
Proton	$1·10 \pm 0·15$	$7·8 \pm 0·8$	18·8
Deuteron	$0·63 \pm 0·03$	$8·0 \pm 0·5$	22·4
Triton	$6·42 \pm 0·20$	$8·0 \pm 0·3$	24·5
^{3}He	$< 7·5 \times 10^{-2}$		
^{4}He	100	$16·0 \pm 0·2$	37·9
^{6}He	$1·95 \pm 0·15$	$12·0 \pm 0·5$	32·6
^{8}He	$(6·2 \pm 0·8) \times 10^{-2}$	$10·2 \pm 1·0$	28·7
Li ions	$0·126 \pm 0·015$	$20·0 \pm 1·0$	38·0
Be ions	$0·156 \pm 0·016$	$\sim 26·0$	45·0

introduced above, and if the transition states were fully understood, then an explanation might be forthcoming within the framework of the theory sketched above.

11.8 Shape isomerism

The adequacy of the above theory was put in doubt by the discovery in 1962 of an isomeric state in ^{242}Am which decayed by spontaneous fission with a half-life of 14 ms. The ground state of ^{242}Am is believed to have a spontaneous-fission of half-life of the order of 10^{10} years. No satisfactory explanation could be given in terms of the theory of the speed at which spontaneous fission proceeded from the isomeric state and why it was so much more favoured than α-decay. Strutinsky (1969) (for further information see L. Wilets, *Theories of Nuclear Fission*, Clarendon Press, 1964) has proposed an elaboration of fission theory which provides an explanation of the behaviour of this isomeric state and of many others which have been subsequently found. It also opens up other theoretical and experimental possibilities.

We recall that the liquid-drop model predicted a smooth variation of nuclear binding energy with A-value, which was in broad agreement with the measured binding energies. This smooth curve was found however to be 'modulated' by effects of shell corrections. In particular, at Z- and N-values corresponding to closed shells we saw that there was a marked increase in binding energies. The shell correction is related to the single-particle level density in the neighbourhood of the highest occupied levels in the single-particle level scheme. The correction is greatest when the level density is low. Thus, for spherical nuclei with closed shells and degenerate levels, the effect is most marked.

It will be appreciated from Figure 61 (p. 170) that the level density determining the shell correction will not only differ from one nucleus to another but will, in the case of a particular nucleus, change with the nuclear deformation. From the Nilsson diagram of Figure 61 it can be seen that the level density can go through a minimum for a deformation greater than the ground-state deformation. Thus the

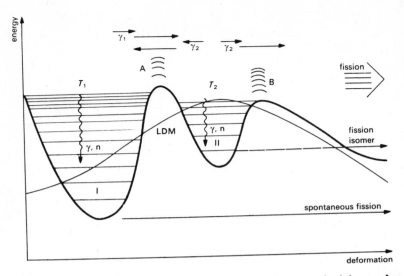

Figure 89 Potential energy against deformation showing second minimum. As
well as excited states within the first well (I), note the transition states at A and
B and the intermediate states in the second well (II). The liquid-drop model
prediction is also shown

shell correction, according to Strutinsky's calculations, can be positive or negative
for the same nucleus depending on the deformation. This has the effect of
modifying the curve of Figure 85 (p. 223) so that it has, as shown in Figure 89, a
second minimum. If this shape has validity then we can associate the isomeric
spontaneous fissioning state with a nucleus deformed beyond the first saddle
point and situated in the second potential minimum. On this view, the
phenomenon has been termed *shape isomerism*. From this isomeric state the
nucleus may return to the first minimum or it may proceed to the scission point,
barrier penetration being involved in both eventualities. If the second well is deep
enough there will then be the possibility of a system of excited states based on
this deformed nucleus. These states are referred to as *intermediate states*.

Many cases of shape isomerism have now been found. In the nucleus ^{240}Pu
there appear to be two such isomeric states, one with a half-life of 30 ns and the
other with a half-life of 5 ns. The existence of the second potential minimum is
thus thought to be a general characteristic of the fissile nuclei.

There is interesting evidence for the existence of intermediate states in the
behaviour of the neutron-capture cross-section in ^{239}Pu at low neutron energies.
Very many resonances are found in neutron capture, only a few of which give
rise to fission. These resonances, shown in Figure 90, which lie below the so-
called 'threshold' for fission, are now interpreted as arising when the position of

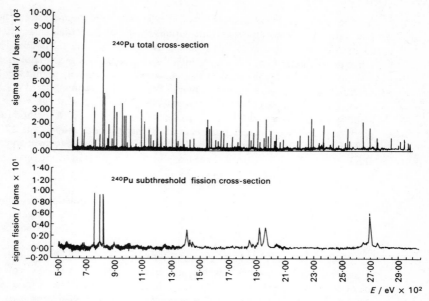

Figure 90 The cross-section for capture of neutrons in ^{240}Pu for all the processes which can occur and the cross-section for capture leading to fission. The grouping of the fission resonances is associated with the intermediate states at the second minimum

the compound nuclear level in the first well lies at the same height as an intermediate level in the second well.

11.9 Superheavy nuclei

These recent theoretical developments have not only provided explanations of previously puzzling aspects of the fission process in the region $Z \simeq 90, A \simeq 240$ but have encouraged speculation about the possible existence of *superheavy nuclei*. As we saw in section 11.4, the liquid-drop model predicts that for nuclei with $Z^2/A > 44$ fission instability should set in. However, shell corrections in uranium and plutonium, neither of which is a magic nucleus, confer greater stability than would have been anticipated. It has therefore been suggested that perhaps in heavier nuclei which are magic or near magic in both proton and neutron numbers, the shell correction can be large enough to restore some measure of stability against spontaneous fission. Following this suggestion we note that the next magic number for protons on the single-particle model would be 126. However, theoretical considerations suggest that there may be large shell corrections in a somewhat lower range of Z-values namely 114 to 120. Above 126 the next single-particle magic number is 184. Values of Z around 114 and N around 184 have then been considered as defining an island on the nuclear chart in which nuclei of short but measurable (or perhaps even quite long) half-life, the so called *superheavy nuclei* might exist. They are expected theoretically to be relatively stable against β-decay. Calculations have been performed in an attempt to arrive at estimates of half-lives against α-decay and spontaneous fission. The fierce dependence of half-life on potential barrier, and the large extrapolation involved in calculating the shell correction, would unfortunately permit any half-life ranging from nanoseconds to millions of years to be accommodated.

The prospect, however nebulous, of extending the region of stability or quasi-stability on the nuclear chart has aroused considerable interest. A method for producing superheavy nuclei in the laboratory that has been proposed is the bombardment of a uranium target by uranium ions accelerated to an energy of about eight million electronvolts per nucleon. This might lead to the reaction

$$^{238}_{92}\text{U} + ^{238}_{92}\text{U} \rightarrow ^{298}_{114}\text{SH} + ^{174}_{70}\text{Yb} + 4\text{n}.$$

The accelerators necessary for such experiments do not at present exist. The prospect of the production of superheavy nuclei is one of the main arguments put forward to justify the development of larger and more sophisticated heavy-ion accelerators.

Should superheavy nuclei be found with lifetimes long enough to permit the measurement of their properties, it is interesting to note that not only would they be of outstanding interest as new forms of nuclear matter but they would open up new fields of atomic physics. The binding energy of the K-electron in uranium, about 114 keV, is still quite small compared to the 1 MeV gap between the positive and negative energy states of the electron. For the superheavy nuclei the K-electrons would lie much deeper, $Z/137$ would be appreciably closer to unity and certain approximations quite satisfactory for known nuclei would have to be reconsidered. If it ever became possible to produce and study nuclei with $Z > 137$ then the electrodynamics of the K-shell would be expected to be very different from that of the nuclei at present known.

11.10   Summary

The challenge of the experimental facts of spontaneous and induced fission has led to a theory for the behaviour of a grossly deformed nucleus. Basically liquid drop in structure, it has had collective states and shell corrections now built into it. A second minimum occurs in the potential energy against deformation curve and indicates how important the shell corrections can be. It has been conjectured that these corrections may be large enough in the neighbourhood of higher magic neutron and proton numbers to lead to superheavy nuclei long enough lived to be identified. Experimental exploration of these suggested 'islands of stability' lying off the known 'peninsula' of stable nuclides awaits the development of accelerators capable of accelerating ions much heavier than those that can be accelerated in presently existing machines.

Chapter 12
Nuclear Reactions

12.1   Introduction

In this chapter we discuss those nuclear reactions, termed 'inelastic', which we define as events in which the nucleons in a nucleus are rearranged by the action of an outside agency. The study of nuclear reactions had its beginnings in experiments performed in 1919 by Lord Rutherford. Nitrogen, bombarded with 7·69 MeV alpha particles from RaC (^{214}Po), was found to eject protons, identifiable as such by their range exceeding that of the initiating alpha particles. Since this pioneer investigation the subject has so expanded that it now requires much more space than is here available for its adequate treatment. In this chapter we limit ourselves to an outline of the fundamentals of nuclear reactions at energies below that necessary for the production of mesons.

12.2   Notation

The following notation is used in the discussion of reactions. We denote the projectile, which constitutes the 'outside agency', by a and the target nuclide by A; assuming that there is a light product b leaving a residual nuclide B, we may write the reaction as

$$a + A \rightarrow b + B \qquad\qquad \textbf{12.1}$$

or more compactly as $A\,(a, b)\,B$.

Thus Rutherford's pioneer experiment may be written as

$$^4\text{He}_2 + {}^{14}\text{N}_7 \rightarrow {}^1\text{H}_1 + {}^{17}\text{O}_8 \text{ or contracted to } {}^{14}\text{N}(\alpha, p)\,{}^{17}\text{O}.$$

The notation can be expanded to accommodate multiple products. For example, if two particles, b and c, are produced, **12.1** becomes

$$a + A \rightarrow b + c + B$$

and the reaction is written $A(a,bc)B$.

12.3   Conservation of nucleon number

In the energy range we are here considering there is not sufficient energy to provide the rest mass involved in nucleon creation and, since nucleons cannot be created, the total number of nucleons is unaltered by the reaction. The production of mesons and other elementary particles is also energetically impossible and so there can be no switch of identity as between neutrons and protons. Therefore not only is the total number of nucleons conserved but the numbers of protons and neutrons are separately conserved. Thus the sum of the A values must be the same on each side of relations like 12.1 and the same condition must be separately satisfied by the Z and N values. This has been assumed in writing 12.1. Note that, within these restrictions, the Z value of the residual nucleus may be different from that of the target; when this is so, transformation of the chemical elements, the ambition of the alchemists, has been realised!

12.4   **Conservation of total energy**

A second quantity which must remain unaltered as a result of the reaction is the total energy of the system. The total energy on one side of the equation is the sum of the rest masses of all the particles plus their kinetic energies, the quantum energy of any electromagnetic photons involved either prior to or following the interaction being added to the kinetic energies. It is customary to introduce Q, as we already did in section 3.2, to denote the decrease in the system rest mass as a result of the reaction. Thus

$$Q = \text{Initial rest mass} - \text{Final rest mass}$$
$$= m_1 + M_1 - (m_2 + M_2) \qquad\qquad 12.$$

where the projectile and target have masses m_1 and M_1, the light and heavy products masses m_2 and M_2 respectively. As the masses enter in differences, we may insert their values as excess masses (section 5.8 and Appendix A) and obtain Q directly in MeV (section 5.8).

The conservation of total energy, where no photons are involved, gives

$$m_1 + M_1 + t_1 + T_1 = m_2 + M_2 + t_2 + T_2$$

where t_1 and T_1 are the kinetic energies of projectile and target and t_2 and T_2 the kinetic energies of the light and heavy products.

We have, therefore,

$$Q = \text{Final kinetic energy} - \text{Initial kinetic energy}$$
$$= t_2 + T_2 - (t_1 + T_1)$$

as an alternative to 12.2.

Q may be positive or negative. If it is positive, kinetic energy is increased as a result of the reaction, which is said to be *exoergic*; if Q is negative, then kinetic energy is absorbed and the reaction is termed *endoergic*. Again, if there is emission of electromagnetic radiation in the course of the reaction, then photon energies have to be included with the kinetic energies.

Note that for exoergic reactions the process is energetically allowed for all bombarding energies, whereas for endoergic reactions the process is energetically forbidden until a certain threshold bombarding energy is reached.

Taking excess mass values for ^{4}He, ^{14}N and ^{1}H from Appendix A and the excess mass of ^{17}O as -0.8077 MeV and substituting in equation 12.2, we find that the original nitrogen experiment is endoergic, with a Q-value of -1.192 MeV.

12.5   Kinematics of two-body reactions

When only two reaction products are involved, the conservation of a third quantity, namely linear momentum, allows us to derive the kinematical relationship between the projectile motion and the motion of the reaction products. Considerable simplification is achieved in the discussion by using the properties of the 'centre of mass' or C-system. As discussed in section 3.6, this is the coordinate system moving with the centre of mass of the projectile and target. In this system, as we shall now show, the projectile and target are approaching each other with equal and oppositely directed linear momenta. As initially the resultant linear momentum in this system is zero, it follows from the conservation requirement that after the interaction the products must go off in the C-system with equal and opposite linear momenta.

Figure 12.1(a) depicts the velocities of projectile and target in the laboratory (or L-system) before the reaction, the target being stationary in the

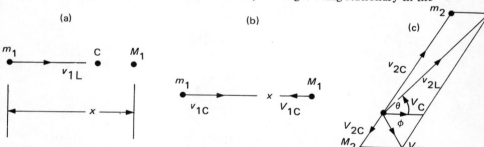

Figure 12.1 Velocity diagrams: (a) L-system, (b) C-system, (c) conversion from C-system to L-system

laboratory. When the projectile to target distance is x, the distance from the target to the centre of mass of these two particles is

$$\frac{m_1 x}{(m_1 + M_1)}$$

The velocity of the centre of mass with respect to the stationary target is therefore

$$V_C = \frac{m_1 v_{1L}}{(m_1 + M_1)} \qquad \qquad 12.3$$

Note that relative to the centre of mass, i.e. in the C-system, the velocity of the projectile is $v_{1L} - V_C$ and hence its linear momentum is $m (v_{1L} - V_C)$, which, using 12.3, is seen to be $M_1 V_C$; relative to the centre of mass, the velocity of the target nucleus is $-V_C$ and so its linear momentum is $-M_1 V_C$. Thus the momentum balance mentioned above is confirmed. Because of this property the C-system is sometimes referred to as the 'zero momentum system'.

We now transfer into the system moving with this velocity, viz. the C-system, by adding $-V_C$ to the velocity of each particle. Figure 12.1(b) illustrates the motion in the C-system. To transfer back to the L-system we add V_C to the velocities in Figure 12.1(b) to get the result shown in Figure 12.1(c).

Note that, assuming the particle velocities concerned to be low enough for non-relativistic dynamics to apply, the initial kinetic energy of the system, which is

$$\frac{m_1 V_1{}^2{}_L}{2}$$

in the L-system, is

$$m_1 \frac{(v_{1L} - V_C)^2}{2} + \frac{M_1 V_C^2}{2}$$

in the C-system. Using 12.3, this can be expressed as

$$\frac{M_o V_{1L}{}^2}{2}$$

where M_o, called the reduced mass, is

$$\frac{m_1 M_1}{(m_1 + M_1)}$$

If the reaction is endoergic there has to be a certain minimum energy input to permit the creation of rest mass-energy; the bombarding energy has thus to exceed this minimum, which is termed the 'threshold energy', if the reaction is to take place. In the C-system, where the products can be produced at rest without violation of linear momentum conservation, the threshold condition is that the initial kinetic energy equals Q in magnitude, Q for the endoergic reaction being negative. Using the above relation between the energies in the two systems, we have that the threshold energy in the L-system is $-m_1 Q/M_o$. The products, produced at rest in the C-system, will both have velocity V_C in the L-system. The energy in excess of $-Q$ provides the necessary forward motion.

If we take the kinematics illustrated in Figure 12.2, which applies to all exoergic reactions and to endoergic reaction above threshold energies, we have from conservation of longitudinal and transverse linear momentum

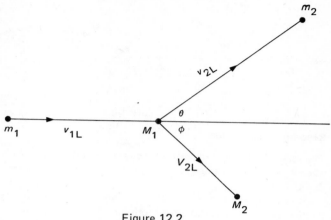

Figure 12.2

$$m_1 v_{1L} = m_2 v_{2L} \cos\theta + M_2 V_{2L} \cos\phi \qquad\qquad 12.4$$

$$m_2 v_{2L} \sin\theta = M_2 V_{2L} \sin\phi \qquad\qquad 12.5$$

and from conservation of mass-energy

$$\frac{m_1 v_{1L}^2}{2} + Q = \frac{m_2 v_{2L}^2}{2} + \frac{M_2 V_{2L}^2}{2} \qquad\qquad 12.6$$

Eliminating ϕ and V_{2L} from equations 12.4, 12.5 and 12.6, we find that

$$Q = \frac{t_2 (M_2 + m_2)}{M_2} - \frac{t_1 (M_2 - m_1)}{M_2} - \frac{2\sqrt{m_1 t_1 m_2 t_2}}{M_2} \cos\theta$$

This equation enables us to find the Q-value of a reaction from a measurement, at a known bombardment energy t_1, of t_2, the energy of the light product travelling at an angle θ in the laboratory system.

In this case the masses enter in ratios, and atomic masses (section 5.3) should be used. However, these are so close to integers that here we may insert the appropriate integer (e.g. 4 for the alpha particle, 1 for the proton and 16 for ^{16}O) without introducing significant error.

We see from Figure 12.3 how, at a fixed angle of observation in the laboratory system, there will be two energy groups of one particular reaction product, corresponding to emission at a forward angle Φ and at a backward angle Φ' in the C-system.

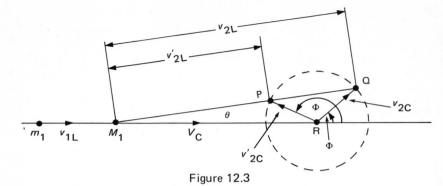

Figure 12.3

12.6 Reaction cross-section

The reaction rate under particular experimental conditions is, in the case of nuclear reactions, discussed in terms of the *reaction cross-section*, which we define as follows.

Suppose a beam of B projectiles per square centimetre per unit time, B being constant across the cross-sectional area of the beam, is incident on a target of area A, thickness t, containing N target nuclei per unit volume of target.

Let dC be the number of reactions per unit time resulting in a product triggering the detector system, assumed to be 100 per cent efficient, of Figure 12.4. Let $d\Omega$ be the solid angle subtended by the detector at the target, assumed to be the same for all points on the target.

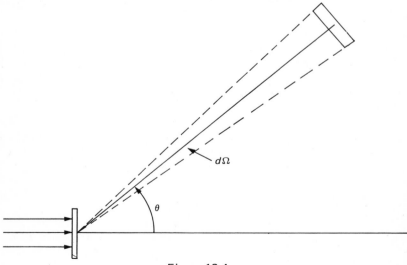

Figure 12.4

dC will be proportional to $B \times N \times t \times A \times d\Omega$, i.e. the counts per unit time per unit solid angle, $dC/d\Omega$, will be proportional to $B \times N \times t \times A$. The constant of proportionality is written $dC/d\Omega$ and is termed the *differential cross-section* at an angle θ. If an integration is performed over all values of θ, we obtain the *total cross-section* σ.

Defined in this way, σ is seen to have the dimensions of area.

The cross-section concept enables us to give a simple picture of the penetration of a projectile through a target containing reaction centres. Imagine each centre to be surrounded by a 'hit' area σ. A reaction is then assumed to take place with that centre if the projectile passes through the 'hit' area; otherwise it is assumed that there is no reaction with that particular centre.

Thus the probability that the projectile will penetrate a thickness of target t is $(1 - N\sigma t)$. This is strictly true only if $N\sigma t \ll 1$. If this inequality is not satisfied, then one target may overlap others and the simple comparison of total target area to total area to give the probability of interaction is not valid. However, if we divide a target of finite thickness into a set of thin slices each of thickness dx we can apply the simple view to each slice. If now we consider a beam of projectiles, B, passing through unit area per unit time, since the probability of any one projectile interacting in the layer thickness dx is $N\sigma dx$, the total number of particles interacting, and hence lost to the beam, will be $dB = -BN\sigma dx$. The solution of the differential equation

$$dB/dx = -BN\sigma \qquad\qquad 12.7$$

viz. $B = B_0 \exp(-N\sigma x)$ describes the exponential absorption of the beam due to the particular nuclear reaction to which σ refers. Note the analogy with equation 2.2 for radioactive decay.

This result may be of practical importance for arriving at the attenuation of a beam of X-rays, γ-rays or neutrons which do not lose energy by ionisation and which in certain circumstances may have a large cross-section for one particular reaction; in the case of ionising particles the steady loss of energy by ionisation dominates, and the above consideration is not directly applicable.

12.7 A general description of the reaction mechanism

We begin by considering the case of a nuclear reaction initiated by a neutron, an uncharged projectile which does not experience the Coulomb force. If the collision impact parameter as defined in section 3.6 is small enough, the neutron will reach the surface of the nucleus and come under the influence of the 'strong' interaction forces. On the assumption that the neutron retains its identity and interacts with an average potential arising from all the nucleons in the target nucleus, the scattering of the projectile can be calculated. This constitutes the independent particle 'optical' model. If the reaction does not proceed beyond this stage and the neutron, deflected by the average potential, simply leaves the nucleus, the neutron is said to have suffered 'shape-elastic scattering'. 'Elastic' implies that the target nucleus is not 'excited' from its initial state as a result of the passage of the neutron but simply recoils in its ground state to conserve linear momentum. Note that on a wave-mechanical view, at certain neutron energies, the neutron wavelength will be such that its wave function will 'resonate' with the potential well by satisfying the boundary conditions at the nuclear surface. Therefore in an experiment devised to measure the total absorption cross-section of a neutron beam passing through a thin foil, shape-elastic scattering should produce resonance peaks in the cross-section versus neutron energy curve.

In many cases, however, the reaction proceeds further, because the neutron, while in the potential well, makes a close collision with a nucleon, or group of nucleons, in the target nucleus. The first opportunity for such a collision will occur as the neutron crosses the surface to enter the target volume. If a struck surface nucleon, or nucleon group, is ejected from the nucleus the event is termed a 'surface direct' reaction. If the neutron crosses the surface without making a collision and then proceeds to collide with and eject interior nucleons, the reaction is termed 'volume direct'.

The optical model can be extended to include these direct interactions by introducing a complex absorption coefficient for the neutron wave. If both the neutron and the struck particle or group of particles come out without further collisions, we have a reaction such as $A(n, np)B$ or $A(n, n)B$. However, either or both may make further collisions, leading to the spread of the kinetic energy brought in by the nucleon (plus the binding energy which has to be added to it) over more and more nucleons. The extension of this multiple collision process to a larger number of particles leads to the formation of a 'compound nucleus' in which the energy is now thoroughly spread over all the nuclear particles. At any one time it may be that none has sufficient energy to leave the nucleus, which may continue in this excited state for a time which is long on the nuclear time scale. Eventually the compound nucleus will decay by statistical fluctuation, leading to one particle having sufficient energy to escape. Figure 12.5 displays the various reaction possibilities against a time scale.

Turning from neutrons to charged projectiles, we have the added complication of the presence of the long-range Coulomb repulsion which the incoming particle has to overcome to get within the range of the strong nuclear force. While a very low-energy neutron can approach the nucleus without hindrance, a charged particle has to have sufficient initial energy to surmount the potential barrier arising from the combined effect of the Coulomb and nuclear forces. This means that the least excitation energy that can be deposited in the compound nucleus is substantially greater than in the case of neutron bombardment. Thus the interesting range of nuclear excitation immediately above the nucleon separation energy where there are many closely spaced levels is not accessible in reactions initiated by charged particles. Also a

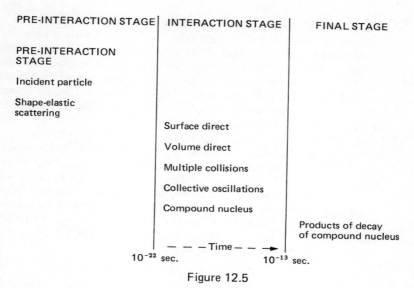

PRE-INTERACTION STAGE	INTERACTION STAGE	FINAL STAGE

PRE-INTERACTION
STAGE

Incident particle

Shape-elastic
scattering

Surface direct

Volume direct

Multiple collisions

Collective oscillations

Compound nucleus

Products of decay
of compound nucleus

$-\ -\ -$ Time $-\ -\ \longrightarrow$

10^{-22} sec. 10^{-13} sec.

Figure 12.5

barrier penetration factor must be included in the theoretical discussion of reactions involving charged particles.

Finally note that if the particle eventually 'boiling off' the compound nucleus is of the same kind as the incident particle, it will be difficult to distinguish compound nucleus formation from shape-elastic scattering. While one process is more prompt than the other, the time interval involved is beyond present techniques to determine. The angular distribution, however, offers the possibility of experimental discrimination.

12.8   Compound nucleus

We conclude this brief treatment of nuclear reactions with some further remarks about the important concept of the compound nucleus, introduced by Neils Bohr in 1936.

The concept arose from the consequences of the observation of very marked resonant behaviour in the absorption cross-section of slow neutrons. While there were resonances to be expected from shape-elastic scattering, these were expected to be broad and widely spaced, corresponding to the fitting of the 'fundamental' and 'harmonics' of the neutron wave function in the fixed-potential well. The observed resonances were closely spaced and their pattern was characteristic of each target nucleus investigated. Moreover they were very narrow, indicating, if the uncertainty principle was to be believed, that there was a comparatively long-lived 'state' involved. Bohr's idea was to associate this long-lived state with the general 'heating' of the whole nucleus described above, achieved by spreading the excitation energy over all the nuclear components. On average, no nucleon in this hot 'liquid drop' had sufficient energy to evaporate. However, there was a comparatively low probability that, by chance collisions, sufficient energy might accumulate on one nucleon, or group of nucleons, such as an alpha particle, and evaporation, involving penetration of the Coulomb barrier in the case of charged products, would result. Thus a physical picture emerged of a state decaying after a time relatively long on a nuclear scale, which could be associated with very narrow observed resonances.

A description of the shape of the resonance is provided by the Breit–Wigner single-level formula, which expresses the probability of the reaction taking place in terms of probability of the compound nucleus being formed multiplied by the probability of its decay. The decay probability enters as a 'decay width', Γ, expressed in MeV, which, by the uncertainty principle, is related to the lifetime of the state. Γ is the 'total width' associated with the probability of decay by any 'exit channel'. Γa and Γb are the widths associated with decay by the particular channels involving the emission of a and b.

From the properties of probabilities it follows that

$$\Gamma = \Gamma a + \Gamma b + \ldots$$

Γa is also associated with probability of entry by the entrance channel, and the Breit–Wigner formula takes the form

$$\sigma ab = \pi \lambda^2 \; g \; \frac{\Gamma a \Gamma b}{[(E - E_0)^2 + \Gamma^2/4]}$$

where σab is the cross-section for the reaction $A(a,b)B$, λ is the wavelength (divided by 2π) of the incident particle in the C-system, E is the energy of excitation of the compound nucleus following capture of the projectile, E_0 is the resonance energy and g, of the order of unity, is a statistical factor depending on the spins.

The Breit–Wigner formula has proved a most successful basis for the theoretical treatment of the compound nucleus. Its formal similarity to the formula for the impedance at resonance of an electrical L,C oscillating circuit incorporating resistance R should be noted. In the nuclear case the decay channels constitute the 'dissipation' which in the electrical case arises from the ohmic losses.

Appendix A
Ground State Properties of Stable Nuclei

	A	Z	N	I	π	SM**	μ‡	Q§	M − A¶	Percentage abundance
n	1	0	1	$\frac{1}{2}$	+		− 1·9131		+ 8·07144	
H	1	1	0	$\frac{1}{2}$	+	$s\frac{1}{2}$	+ 2·79278		+ 7·28899	99·985
	2	1	1	1	+	$(\frac{1}{2},\frac{1}{2})$	+ 0·85742	+ 0·0028	+ 13·136	0·015
He	3	2	1	$\frac{1}{2}$	+	$s\frac{1}{2}$	− 2·1276		+ 14·931	0·00013
	4	2	2	0	+				+ 2·425	100
Li	6	3	3	1	+	$(\frac{3}{2},\frac{3}{2})$	+ 0·82202	− 0·0008	+ 14·088	7·4
	7	3	4	$\frac{3}{2}$	−	$p\frac{3}{2}$	+ 3·2564	− 0·04	+ 14·907	92·6
Be	9	4	5	$\frac{3}{2}$	−	$p\frac{3}{2}$	− 1·1776	+ 0·05	+ 11·350	100
B	10	5	5	3	+	$(\frac{3}{2},\frac{3}{2})$	+ 1·8007	+ 0·08	+ 12·052	18·8
	11	5	6	$\frac{3}{2}$	−	$p\frac{3}{2}$	+ 2·6885	+ 0·04	+ 8·668	81·2
C	12	6	6	0	+				0 (standard)	98·89
	13	6	7	$\frac{1}{2}$	−	$p\frac{1}{2}$	+ 0·7024		+ 3·125	1·11
N	14	7	7	1	+	$(\frac{1}{2},\frac{1}{2})$	+ 0·4036	+ 0·01	+ 2·864	99·63
	15	7	8	$\frac{1}{2}$	−	$p\frac{1}{2}$	− 0·2831		+ 0·100	0·37
O	16	8	8	0	+				− 4·737	99·759
	17	8	9	$\frac{5}{2}$	+	$d\frac{5}{2}$	− 1·8937	− 0·026	− 0·808	0·037
	18	8	10	0	+				− 0·782	0·204
F	19	9	10	$\frac{1}{2}$	+		+ 2·6288		− 1·486	100
Ne	20	10	10	0	+				− 7·041	90·8
	†21	10	11	$\frac{3}{2}$	+	$(d\frac{5}{2})^3$	− 0·6618	+ 0·09	− 5·730	0·26
	22	10	12	0	+				− 8·025	8·9
Na	†23	11	12	$\frac{3}{2}$	+	$(d\frac{5}{2})^3$	+ 2·2175	+ 0·14	− 9·528	100
Mg	24	12	12	0	+				− 13·933	78·8
	25	12	13	$\frac{5}{2}$	+	$d\frac{5}{2}$	− 0·8551	+ 0·22	− 13·191	10·1
	26	12	14	0	+				− 16·214	11·1

** Shell-model configurations.
‡ Nuclear magnetic dipole moment in nuclear magnetons.
§ Nuclear electric quadrupole moment in barns.
¶ Mass excess (see section 5.8) in MeV.

	A	Z	N	I	π	SM	μ	Q	M − A	Percentage abundance
Al	27	13	14	$\frac{5}{2}$	+	$d\frac{5}{2}$	+ 3·6414	+ 0·15	− 17·196	100
Si	28	14	14	0	+				− 21·490	92·17
	29	14	15	$\frac{1}{2}$	+	$s\frac{1}{2}$	− 0·5553		− 21·894	4·71
	30	14	16	0	+				− 24·439	3·12
P	31	15	16	$\frac{1}{2}$	+	$s\frac{1}{2}$	+ 1·1317		− 24·438	100
S	32	16	16	0	+				− 26·013	95
	33	16	17	$\frac{3}{2}$	+	$d\frac{3}{2}$	+ 0·6433	− 0·055	− 26·583	0·75
	34	16	18	0	+				− 29·933	4·2
	36	16	20	0	+				− 30·655	0·017
Cl	35	17	18	$\frac{3}{2}$	+	$d\frac{3}{2}$	+ 0·82183	− 0·079	− 29·014	75·53
	37	17	20	$\frac{3}{2}$	+	$d\frac{3}{2}$	+ 0·68411	− 0·062	− 31·765	24·47
A	36	18	18	0	+				− 30·232	0·337
	38	18	20	0	+				− 34·718	0·063
	40	18	22	0	+				− 35·038	99·60
K	39	19	20	$\frac{3}{2}$	+	$d\frac{3}{2}$	+ 0·3914	+ 0·055	− 33·803	93·2
	*40	19	21	4			− 1·298	− 0·07	− 33·533	0·0119
	41	19	22	$\frac{3}{2}$	+	$d\frac{3}{2}$	+ 0·2149	+ 0·067	− 35·552	6·8
Ca	40	20	20	0	+				− 34·848	96·9
	42	20	22	(0)	(+)				− 38·540	0·64
	43	20	23	$\frac{7}{2}$	−	$f\frac{7}{2}$	− 1·317		− 38·396	0·14
	44	20	24	(0)	(+)				− 41·460	2·1
	46	20	26	(0)	(+)				− 43·138	**0·0032**
	48	20	28	(0)	(+)				− 44·216	0·18
Sc	45	21	24	$\frac{7}{2}$	−	$f\frac{7}{2}$	+ 4·7564	− 0·22	− 41·061	100
Ti	46	22	24	(0)	(+)				− 44·123	8·0
	†47	22	25	$\frac{5}{2}$	−	$(f\frac{7}{2})^5$	− 0·7883	+ 0·29	− 44·927	7·4
	48	22	26	(0)	(+)				− 48·483	73·8
	49	22	27	$\frac{7}{2}$	−	$f\frac{7}{2}$	− 1·1039	+ 0·24	− 48·558	5·5
	50	22	28	(0)	(+)				− 51·431	5·3
V	*50	23	27	6	+		+ 3·3470	± 0·06	− 49·216	0·25
	51	23	28	$\frac{7}{2}$	−	$f\frac{7}{2}$	+ 5·149	− 0·05	− 52·199	99·75
Cr	50	24	26	(0)	(+)				− 50·249	4·4
	52	24	28	(0)	(+)				− 55·411	83·7
	53	24	29	$\frac{3}{2}$	−	$p\frac{3}{2}$	− 0·4744	± 0·03	− 55·281	9·5
	54	24	30	(0)	(+)				− 56·930	2·4
Mn	†55	25	30	$\frac{5}{2}$	(−)	$(f\frac{7}{2})^5$	+ 3·444	+ 0·4	− 57·705	100
Fe	54	26	28	(0)	(+)				− 56·245	5·9
	56	26	30	(0)	(+)				− 60·605	91·6
	57	26	31	$\frac{1}{2}$	−	$p\frac{1}{2}$	+ 0·0902		− 60·175	2·20
	58	26	32	(0)	(+)				− 62·146	0·33

	A	Z	N	I	π	SM	μ	Q	M − A	Percentage abundance
Co	59	27	32	$\frac{7}{2}$	−	$f_{\frac{7}{2}}$	+4·62	+0·4	−62·233	100
Ni	58	28	30	(0)	(+)				−60·228	68·0
	60	28	32	(0)	(+)				−64·471	26·2
	61	28	33	$\frac{3}{2}$	−	$p_{\frac{3}{2}}$	−0·7487	+0·16	−64·220	1·1
	62	28	34	(0)	(+)				−66·748	3·7
	64	28	36	(0)	(+)				−67·106	1·0
Cu	63	29	34	$\frac{3}{2}$	−	$p_{\frac{3}{2}}$	+2·223	−0·180	−65·583	69·0
	65	29	36	$\frac{3}{2}$	−	$p_{\frac{3}{2}}$	+2·382	−0·195	−67·266	31·0
Zn	64	30	34	0	+				−66·000	48·9
	66	30	36	0	+				−68·881	27·8
	67	30	37	$\frac{5}{2}$	−	$f_{\frac{5}{2}}$	+0·8754	+0·17	−67·863	4·1
	68	30	38	0	+				−69·994	18·6
	70	30	40	(0)	(+)				−69·550	0·63
Ga	69	31	38	$\frac{3}{2}$	−	$p_{\frac{3}{2}}$	+2·016	+0·19	−69·326	60·1
	71	31	40	$\frac{3}{2}$	−	$p_{\frac{3}{2}}$	+2·562	+0·12	−70·135	39·9
Ge	70	32	38	0	+				−70·558	20·5
	72	32	40	0	+				−72·579	27·4
	73	32	41	$\frac{9}{2}$	+	$g_{\frac{9}{2}}$	−0·8792	−0·28	−71·293	7·8
	74	32	42	0	+				−73·418	36·5
	76	32	44	0	+				−73·209	7·8
As	75	33	42	$\frac{3}{2}$	−	$p_{\frac{3}{2}}$	+1·439	+0·29	−73·031	100
Se	74	34	40	0	+				−72·212	0·93
	76	34	42	0	+				−75·257	9·1
	77	34	43	$\frac{1}{2}$	−	$p_{\frac{1}{2}}$	+0·534		−74·601	7·5
	78	34	44	0	+				−77·020	23·6
	80	34	46	0	+				−77·753	49·9
	82	34	48	0	+				−77·586	9·0
Br	79	35	44	$\frac{3}{2}$	−	$p_{\frac{3}{2}}$	+2·106	+0·31	−76·075	50·6
	81	35	46	$\frac{3}{2}$	−	$p_{\frac{3}{2}}$	+2·270	+0·26	−77·972	49·4
Kr	78	36	42	(0)	(+)				−74·143	0·35
	80	36	44	(0)	(+)				−77·891	2·27
	82	36	46	0	+				−80·589	11·6
	83	36	47	$\frac{9}{2}$	+	$g_{\frac{9}{2}}$	−0·970	+0·26	−79·985	11·5
	84	36	48	0	+				−82·433	57·0
	86	36	50	0	+				−83·259	17·3
Rb	85	37	48	$\frac{5}{2}$	−	$f_{\frac{5}{2}}$	+1·3524	+0·26	−82·156	72·2
	*87	37	50	$\frac{3}{2}$	−	$p_{\frac{3}{2}}$	+2·7500	+0·12	−84·591	27·8

	A	Z	N	I	π	SM	μ	Q	M − A	Percentage abundance
Sr	84	38	46	(0)	(+)				− 80·638	0·55
	86	38	48	0	+				− 84·499	9·8
	87	38	49	$\frac{9}{2}$	+	$g\frac{9}{2}$	− 1·093	+ 0·30	− 84·865	7·0
	88	38	50	0	+				− 87·894	82·7
Y	89	39	50	$\frac{1}{2}$	−	$p\frac{1}{2}$	− 0·1373		− 87·678	100
Zr	90	40	50	(0)	(+)				− 88·770	51·5
	91	40	51	$\frac{5}{2}$	+	$d\frac{5}{2}$	− 1·303		− 87·893	11·2
	92	40	52	(0)	(+)				− 88·462	17·1
	94	40	54	(0)	(+)				− 87·267	17·4
	96	40	56	(0)	(+)				− 85·430	2·8
Nb	93	41	52	$\frac{9}{2}$	+	$g\frac{9}{2}$	+ 6·167	− 0·22	− 87·203	100
Mo	92	42	50	0	+				− 86·804	15·7
	94	42	52	0	+				− 88·406	9·3
	95	42	53	$\frac{5}{2}$	+	$d\frac{5}{2}$	− 0·9133	± 0·12	− 87·709	15·7
	96	42	54	0	+				− 88·794	16·5
	97	42	55	$\frac{5}{2}$	+	$d\frac{5}{2}$	− 0·9325	± 1·1	− 87·539	9·5
	98	42	56	0	+				− 88·110	23·8
	100	42	58	0	+				− 86·185	9·5
Tc		43	no stable isotope							
Ru	96	44	52	(0)	(+)				− 86·071	5·6
	98	44	54	(0)	(+)				− 88·221	1·9
	99	44	55	$\frac{5}{2}$	+	$d\frac{5}{2}$	− 0·63		− 87·619	12·7
	100	44	56	(0)	(+)				− 89·219	12·7
	101	44	57	$\frac{5}{2}$	+	$d\frac{5}{2}$	− 0·69		− 87·953	17·0
	102	44	58	(0)	(+)				− 89·098	31·5
	104	44	60	(0)	(+)				− 88·090	18·6
Rh	103	45	58	$\frac{1}{2}$	−	$p\frac{1}{2}$	− 0·0883		− 88·014	100
Pd	102	46	56	(0)	(+)				− 87·923	1·0
	104	46	58	(0)	(+)				− 89·411	11·0
	105	46	59	$\frac{5}{2}$	+	$d\frac{5}{2}$	− 0·642	+ 0·8	− 88·431	22·2
	106	46	60	(0)	(+)				− 89·907	27·3
	108	46	62	(0)	(+)				− 89·524	26·7
	110	46	64	(0)	(+)				− 88·338	11·8
Ag	107	47	60	$\frac{1}{2}$	−	$p\frac{1}{2}$	− 0·1135		− 88·403	51·4
	109	47	62	$\frac{1}{2}$	−	$p\frac{1}{2}$	− 0·1305		− 88·717	48·6
Cd	106	48	58	(0)	(+)				− 87·128	1·22
	108	48	60	(0)	(+)				− 89·248	0·88
	110	48	62	0	+				− 90·342	12·4
	111	48	63	$\frac{1}{2}$	+	$s\frac{1}{2}$	− 0·5943		− 89·246	12·8
	112	48	64	0	+				− 90·575	24·0
	113	48	65	$\frac{1}{2}$	+	$s\frac{1}{2}$	− 0·6217		− 89·041	12·3

	A	Z	N	I	π	SM	μ	Q	M − A	Percentage abundance
Cd	114	48	66	0	+				− 90·018	28·8
	116	48	68	0	+				− 88·712	7·6
In	113	49	64	$\frac{9}{2}$	+	g^{9}_{2}	+ 5·523	+ 0·82	− 89·339	4·2
	*115	49	66	$\frac{9}{2}$	+	g^{9}_{2}	+ 5·534	+ 0·83	− 89·542	95·8
Sn	112	50	62	(0)	(+)				− 88·644	1·02
	114	50	64	(0)	(+)				− 90·565	0·69
	115	50	65	$\frac{1}{2}$	+	s^{1}_{2}	− 0·918		− 90·031	0·38
	116	50	66	0	+				− 91·523	14·3
	117	50	67	$\frac{1}{2}$	+	s^{1}_{2}	− 1·000		− 90·392	7·6
	118	50	68	0	+				− 91·652	24·1
	119	50	69	$\frac{1}{2}$	+	s^{1}_{2}	− 1·046		− 90·062	8·5
	120	50	70	0	+				− 91·100	32·5
	122	50	72	(0)	(+)				− 89·942	4·8
	124	50	74	(0)	(+)				− 88·237	6·1
Sb	121	51	70	$\frac{5}{2}$	+	d^{5}_{2}	+ 3·359	− 0·29	− 89·593	57
	123	51	72	$\frac{7}{2}$	+	g^{5}_{2}	+ 2·547	− 0·37	− 89·224	43
Te	120	52	68	(0)	(+)				− 89·400	0·091
	122	52	70	(0)	(+)				− 90·291	2·5
	*123	52	71	$\frac{1}{2}$	+	s^{1}_{2}	− 0·7359		− 89·163	0·88
	124	52	72	(0)	(+)				− 90·500	4·6
	125	52	73	$\frac{1}{2}$	+	s^{1}_{2}	− 0·8871		− 89·032	7·0
	126	52	74	0	+				− 90·053	18·7
	128	52	76	0	+				− 88·978	31·8
	130	52	78	0	+				− 87·337	34·4
I	127	53	74	$\frac{5}{2}$	+	d^{5}_{2}	+ 2·808	− 0·79	− 88·984	100
Xe	124	54	70	(0)	(+)				− 87·450	0·094
	126	54	72	(0)	(+)				− 89·154	0·092
	128	54	74	(0)	(+)				− 89·850	1·92
	129	54	75	$\frac{1}{2}$	+	s^{1}_{2}	− 0·7768		− 88·692	26·4
	130	54	76	(0)	(+)				− 89·880	4·1
	131	54	77	$\frac{3}{2}$	+	d^{3}_{2}	+ 0·6908	− 0·12	− 88·411	21·2
	132	54	78	0	+				− 89·272	26·9
	134	54	80	0	+				− 88·120	10·4
	136	54	82	0	+				− 86·422	8·9
Cs	133	55	78	$\frac{7}{2}$	+	g^{7}_{2}	+ 2·578	− 0·003	− 88·160	100
Ba	130	56	74	(0)	(+)				− 87·331	0·101
	132	56	76	(0)	(+)				− 88·380	0·097
	134	56	78	0	+				− 88·852	2·42
	135	56	79	$\frac{3}{2}$	+	d^{3}_{2}	+ 0·8365	+ 0·18	− 87·980	6·6
	136	56	80	0	+				− 89·140	7·8
	137	56	81	$\frac{3}{2}$	+	d^{3}_{2}	+ 0·9357	+ 0·28	− 88·020	11·3
	138	56	82	0	+				− 88·490	71·7

	A	Z	N	I	π	SM	μ	Q	M − A	Percentage abundance
La	*138	57	81	5	−		+ 3·707	± 0·8	− 86·710	0·089
	139	57	82	$\frac{7}{2}$	+	$g\frac{7}{2}$	+ 2·778	+ 0·22	− 87·428	99·911
Ce	136	58	78	(0)	(+)				− 86·550	0·19
	138	58	80	(0)	(+)				− 87·720	0·26
	140	58	82	(0)	(+)				− 88·125	88·47
	142	58	84	(0)	(+)				− 84·631	11·08
Pr	141	59	82	$\frac{5}{2}$	+	$d\frac{5}{2}$	+ 4·3	− 0·07	− 86·072	100
Nd	142	60	82	(0)	(+)				− 86·010	27·1
	143	60	83	$\frac{7}{2}$	−	$f\frac{7}{2}$	− 1·08	− 0·48	− 84·039	12·2
	*144	60	84	(0)	(+)				− 83·797	23·9
	145	60	85	$\frac{7}{2}$	−	$f\frac{7}{2}$	−0·66	− 0·25	− 81·469	8·3
	146	60	86	(0)	(+)				− 80·959	17·2
	148	60	88	(0)	(+)				− 77·435	5·7
	150	60	90	(0)	(+)				− 73·666	5·6
Pm		61	no stable isotope							
Sm	144	62	82	(0)	(+)				− 81·980	3·1
	*147	62	85	$\frac{7}{2}$	−	$f\frac{7}{2}$	− 0·813	− 0·20	− 79·300	15·0
	148	62	86	(0)	(+)				− 79·371	11·2
	149	62	87	$\frac{7}{2}$	−	$f\frac{7}{2}$	− 0·670	+ 0·058	− 77·145	13·8
	150	62	88	(0)	(+)				− 77·056	7·4
	152	62	90	(0)	(+)				− 74·746	26·8
	154	62	92	(0)	(+)				− 72·393	22·7
Eu	151	63	88	$\frac{5}{2}$	(+)	$(d\frac{5}{2})$	+ 3·464	+ 1·1	− 74·670	47·8
	153	63	90	$\frac{5}{2}$	+	$d\frac{5}{2}$	+ 1·530	+ 2·8	− 73·361	52·2
Gd	152	64	88	(0)	(+)				− 74·710	0·20
	154	64	90	(0)	(+)				− 73·653	2·15
	155	64	91	$\frac{3}{2}$	−	$p\frac{3}{2}$	− 0·254	+ 1·3	− 72·037	14·7
	156	64	92	(0)	(+)				− 72·493	20·5
	157	64	93	$\frac{3}{2}$			− 0·339	+ 1·7	− 70·769	15·7
	158	64	94	(0)	(+)				− 70·627	24·9
	160	64	96	(0)	(+)				− 67·891	21·9
Tb	159	65	94	$\frac{3}{2}$	+	$d\frac{3}{2}$	± 1·99	+ 1·3	− 69·534	100
Dy	156	66	90	(0)	(+)				− 70·860	0·052
	158	66	92	(0)	(+)				− 70·374	0·090
	160	66	94	(0)	(+)				− 69·673	2·29
	161	66	95	$\frac{5}{2}$	+	$d\frac{5}{2}$	− 0·46	+ 2·3	− 68·049	18·9
	162	66	96	(0)	(+)				− 68·182	25·5
	163	66	97	$\frac{5}{2}$	−		+ 0·64	+ 2·5	− 66·363	25·0
	164	66	98	(0)	(+)				− 65·949	28·2
Ho	165	67	98	$\frac{7}{2}$	+	$g\frac{7}{2}$	+ 4·12	+ 3·0	− 64·811	100

	A	Z	N	I	π	SM	μ	Q	M − A	Percentage abundance
Er	162	68	94	(0)	(+)				− 66·370	0·136
	164	68	96	(0)	(+)				− 65·867	1·56
	166	68	98	(0)	(+)				− 64·918	33·4
	167	68	99	$\frac{7}{2}$	(−)	$(f^7_{\frac{7}{2}})$	− 0·564	+ 2·8	− 63·285	22·9
	168	68	100	(0)	(+)				− 62·983	27·1
	170	68	102	(0)	(+)				− 60·020	14·9
Tm	169	69	100	$\frac{1}{2}$	(+)	$(s_{\frac{1}{2}})$	− 0·232		− 61·249	100
Yb	168	70	98	(0)	(+)				− 61·330	0·14
	170	70	100	(0)	(+)				− 60·530	3·03
	171	70	101	$\frac{1}{2}$	−	$p_{\frac{1}{2}}$	+ 0·4919		− 59·220	14·3
	172	70	102	(0)	(+)				− 59·280	21·8
	173	70	103	$\frac{5}{2}$	−	$f_{\frac{5}{2}}$	− 0·6776	+ 3·0	− 57·690	16·2
	174	70	104	(0)	(+)				− 57·060	31·8
	176	70	106	(0)	(+)				− 53·390	12·7
Lu	175	71	104	$\frac{7}{2}$	+	$g_{\frac{7}{2}}$	+ 2·23	+ 5·6	− 55·290	97·40
	*176	71	105	7	−		+ 3·18	+ 8·0	− 53·410	2·60
Hf	174	72	102	(0)	(+)				− 55·550	0·18
	176	72	104	(0)	(+)				− 54·430	5·2
	177	72	105	$\frac{7}{2}$	−		+ 0·61	+ 3	− 52·720	18·5
	178	72	106	0	+				− 52·270	27·1
	179	72	107	$\frac{9}{2}$	+		− 0·47	+ 3	− 50·270	13·8
	180	72	108	0	+				− 49·530	35·2
Ta	*180	73	107						− 48·862	0·012
	181	73	108	$\frac{7}{2}$	+	$g_{\frac{7}{2}}$	+ 2·36	+ 4·2	− 48·430	99·988
W	180	74	106	(0)	(+)				− 49·365	0·14
	182	74	108	0	+				− 48·156	26·2
	183	74	109	$\frac{1}{2}$	−	$p_{\frac{1}{2}}$	+ 0·117		− 46·272	14·3
	184	74	110	0	+				− 45·619	30·7
	186	74	112	0	+				− 42·438	28·7
Re	185	75	110	$\frac{5}{2}$	+	$d_{\frac{5}{2}}$	+ 3·172	+ 2·7	− 43·725	37·1
	*187	75	112	$\frac{5}{2}$	+	$d_{\frac{5}{2}}$	+ 3·204	+ 2·6	− 41·140	62·9
Os	184	76	108	(0)	(+)				− 44·010	0·018
	186	76	110	(0)	(+)				− 42·970	1·59
	187	76	111	$\frac{1}{2}$	−	$p_{\frac{1}{2}}$	+ 0·0643		− 41·141	1·64
	188	76	112	(0)	(+)				− 40·909	13·3
	189	76	113	$\frac{3}{2}$	−	$p_{\frac{3}{2}}$	+ 0·6566	+ 0·8	− 38·840	16·1
	190	76	114	(0)	(+)				− 38·540	26·4
	192	76	116	(0)	(+)				− 35·910	41·0
Ir	191	77	114	$\frac{3}{2}$	+	$d_{\frac{3}{2}}$	+ 0·145	+ 1·3	− 36·670	38·5
	193	77	116	$\frac{3}{2}$	+	$d_{\frac{3}{2}}$	+ 0·158	+ 1·2	− 34·454	61·5

	A	Z	N	I	π	SM	μ	Q	M − A	Percentage abundance
Pt	192	78	114	(0)	(+)				− 36·190	0·78
	194	78	116	0	+				− 34·721	32·8
	195	78	117	$\frac{1}{2}$	−	$p\frac{1}{2}$	+0·6060		− 32·776	33·7
	196	78	118	0	+				− 32·633	25·4
	198	78	120	(0)	(+)				− 29·905	7·2
Au	197	79	118	$\frac{3}{2}$	+	$d\frac{3}{2}$	+ 0·144 86	+ 0·58	− 31·166	100
Hg	196	80	116	(0)	(+)				− 31·838	0·15
	198	80	118	0	+				− 30·966	10·0
	199	80	119	$\frac{1}{2}$	−	$p\frac{1}{2}$	+ 0·5027		− 29·547	16·9
	200	80	120	0	+				− 29·503	23·1
	201	80	121	$\frac{3}{2}$	−	$p\frac{3}{2}$	− 0·5567	+ 0·45	− 27·658	13·2
	202	80	122	0	+				− 27·346	29·8
	204	80	124	0	+				− 24·689	6·8
Tl	203	81	122	$\frac{1}{2}$	+	$s\frac{1}{2}$	+ 1·6115		− 25·753	29·5
	205	81	124	$\frac{1}{2}$	+	$s\frac{1}{2}$	+ 1·6274		− 23·807	70·5
Pb	204	82	122	(0)	(+)				− 25·109	1·3
	206	82	124	0	+				− 23·783	26
	207	82	125	$\frac{1}{2}$	−	$p\frac{1}{2}$	+ 0·5895		− 22·446	21
	208	82	126	0	+				− 21·750	52
Bi	209	83	126	$\frac{9}{2}$	−	$h\frac{9}{2}$	+ 4·080	− 0·35	− 18·262	100

*Nuclides which are now known to be unstable. Their half-life is very long and they have traditionally been included in the estimates of percentage abundance.

† Nuclides which depart in respect of spin from 'single-particle' prediction by virtue of having several 'unpaired' nucleons.

[Based on Appendix G, R. B. Leighton, *Principles of Modern Physics*, McGraw-Hill, 1959; W. D. Myers and W. J. Swiatecki, *Nuclear Masses and Deformations*, University of California, UCRL-11980, 1965; G. H. Fuller and V. W. Cohen, *Nuclear Spins and Moments*, Nuclear Data Tables, Academic Press, 1969.]

Appendix B
The Passage of Charged Particles through Matter

An understanding of the processes by which charged particles lose energy when passing through matter, be it a vacuum window or the material of a detector, is obviously essential in the design and interpretation of experiments in nuclear physics. Furthermore, in the early investigations in the field great reliance was placed on range measurements as a means of determining the energy of particles, and consequently the subject of this appendix attracted much theoretical and experimental attention. There has been a revival of interest in this topic with the advent of DSAM (see section 10.12).

The problem is one of great complexity. The particle encounters not only a system of orbital electrons but also the nuclear system and it may interact with either or with both depending on its energy at any particular point in its trajectory. In the case of heavy particles, consisting of one or more nucleons, they may in the course of their passage through the material pick up electrons with a consequent change in their effective charge and their subsequent behaviour. A full treatment of the problem is to be found in R. D. Evans, *The Atomic Nucleus*, McGraw-Hill, 1965. Here only a brief outline will be given of those aspects of relevance to the behaviour of α-particles of energies up to about five million electronvolts (and other light nuclei having comparable velocity), together with aspects of relevance to electrons in the energy range encountered in β-decay, that is up to a few million electronvolts.

B.1 Heavy particles

The following assumptions, which can be considered valid for the heavy particles here of interest, enable us to proceed to a useful semiclassical theory.

(a) The particle pursues a straight path through the material. This neglects the occasional close encounter with a nucleus leading to large-angle scattering.

(b) The main energy loss is by the ionization and excitation of atoms. Energy loss by radiation (bremsstrahlung), or by nuclear interactions via the specifically nuclear forces, is neglected

(c) The particle velocity V is large compared to the orbital velocity u of any electron to which energy transfer takes place. The orbital electrons are therefore regarded as stationary for the purpose of our simple analysis.

(d) V is small compared to c, the velocity of light, so that the treatment can be non-relativistic.

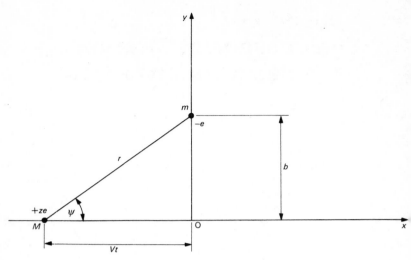

Figure 91 Heavy particle, charge $+ze$, interacting and communicating energy to atomic electron

Consider now the impulse communicated to an electron situated a distance b off the particle's trajectory, as depicted in Figure 91. If the particle's energy loss is small, its velocity will be little affected by the energy transfer to the single electron. We assume that the x-component of the impulse is zero, the effect of the particle in its approach being balanced by its opposite effect as it departs. Measuring t from the instant the particle passes through the origin, we can write p_y, the total impulse communicated, as

$$p_y = \int \text{force } dt$$

$$= \int_{-\infty}^{+\infty} \frac{ze^2}{r^2} \sin \psi \, dt$$

$$= \int_{0}^{\pi} \frac{ze^2 \sin \psi \csc^2 \psi}{b^2 \csc^2 \psi} \frac{b}{V} \, d\psi$$

$$= \frac{2ze^2}{bV}.$$

Now T_e, the electron kinetic energy, will be related to its momentum by the non-relativistic relation

$$T_e = \frac{p_y^2}{2m},$$

where m is the electron mass.

Therefore
$$T_e = \frac{2z^2 e^4}{b^2 m V^2}.$$ **B.1**

Let us now consider the particle traversing a thin foil, as in Figure 92. Let the foil thickness be dx, the charge and mass numbers Z and A. The number of atoms per unit volume of the foil will then be $\rho N_A / A$, where ρ is the density of the foil material and N_A is Avogadro's constant. Consequently the number of

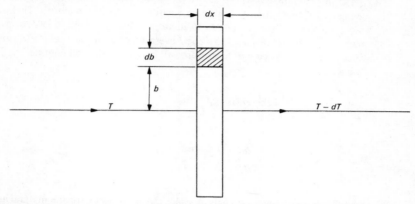

Figure 92 Passage of particle with loss of energy $-dT$ through a thin absorber

electrons per unit volume of the foil will be $Z\rho N_A / A$. The number of electrons lying at a distance between b and $b + db$ from the particle's path will then be

$$\frac{Z\rho N_A}{A} 2\pi b \, db \, dx.$$

To these electrons the particle will therefore lose energy

$$\frac{Z\rho N_A}{A} 2\pi b \, db \, dx \, \frac{2z^2 e^4}{b^2 m V^2}.$$

The total loss of energy of the particle to electrons at all distances in passing through the foil, which we will denote by $-dT$, will therefore be given by

$$-dT = \left[\int_0^\infty \frac{Z\rho N_A}{A} 2\pi b \, \frac{2z^2 e^4}{b^2 m V^2} \, db \right] dx$$

$$= \frac{4\pi Z\rho N_A}{A} \frac{z^2 e^4}{m V^2} \, dx \int_{b_1}^{b_2} \frac{1}{b} \, db.$$ **B.2**

We have now to consider the possible infinity troubles at both limits if we contemplate small values of b_1 and large values of b_2. Taking the lower limit, where our assumption about the transverse impulse will lose validity, we note that in a head-on collision the energy communicated to the electron is, by the conservation of energy and momentum, $\frac{1}{2}m(2V)^2$, assuming $m/M \ll 1$. If we then choose the lowest value of b which we are going to consider to correspond to this maximum energy transfer we have, from equation **B.1**,

$$\tfrac{1}{2}m(2V)^2 = \frac{2z^2e^4}{b_1^2\, mV^2}, \quad \text{i.e.} \quad b_1 = \frac{ze^2}{mV^2}.$$

The upper limit to b will be set by the lowest energy which the electron, having regard to its incorporation in a quantized system, can absorb. This will vary from one electron to another in an atom as it depends on the electron quantum numbers. We assume a value I averaged over all the atomic electrons which, except in the case of hydrogen, will have to be derived empirically. Then we have

$$\frac{2z^2e^4}{b_2^2\, mV^2} = I, \quad \text{and therefore} \quad b_2 = \frac{ze^2}{V}\sqrt{\left[\frac{2}{mI}\right]}.$$

Hence
$$\int_{b_1}^{b_2} \frac{1}{b}\, db = \ln\frac{b_2}{b_1} = \frac{1}{2}\ln\frac{2mV^2}{I}.$$

An exact wave-mechanical analysis yields twice this value for the integral and we include this factor of two in what follows.

With the choice of the above values for the limits of integration, equation **B.2** can be written

$$\left[-\frac{dT}{dx}\right]_{\text{ion}} = \frac{4\pi Z\rho N_A}{A}\frac{z^2e^4}{mV^2}\ln\frac{2mV^2}{I},$$

giving the rate at which energy is lost as a function of the absorber properties and the particle properties including the particle energy. For heavy particles which travel in straight lines, the concept of range, defined to be the maximum amount of material penetrated, is useful. If we denote the range by R, we have

$$R = \int_0^R dx = \int_0^{T_0} \frac{1}{-dT/dx}\, dT,$$

where T_0 is the initial kinetic energy of the particle. Now $T = \frac{1}{2}MV^2$ and hence $dT = MV\, dV$.

Therefore
$$R = \frac{m}{e^4 4\pi N_A}\frac{A}{\rho Z}\frac{M}{z^2}\int_0^{V_0} \frac{V^3}{\ln(2mV^2/I)}\, dV. \qquad \textbf{B.3}$$

Without tackling the integral, which presents some difficulties, we can make some useful deductions concerning (a) different types of particles passing through the same absorbing material and (b) the behaviour of the same particles passing through different absorbers.

If we take particles, masses M_1 and M_2, charge numbers z_1 and z_2, which have the same initial velocity V_0, then their ranges R_1 and R_2 in the same absorber will be in the ratio

$$\frac{R_1}{R_2} = \frac{M_1}{M_2}\left[\frac{z_2}{z_1}\right]^2$$

Their energies T_1 and T_2 will be in the ratio

$$\frac{T_1}{T_2} = \frac{M_1}{M_2}.$$

Thus a deuteron will have twice the range of a proton of half the energy, since $M_1/M_2 = 2$ and $z_2/z_1 = 1$. An α-particle will have the same range as a proton of one quarter of its energy since $M_1/M_2 = 4$, $z_2/z_1 = \frac{1}{2}$.

These simple rules are sometimes of use in extending range–energy relationships from one type of light particle to another. They are subject to serious error in the low-energy region, where electrons may become attached to the particle. The probability of picking up an electron becomes appreciable when the particle's velocity is of the order of or less than the electron orbital velocity, $zc/137$. For protons this velocity is 0.22×10^7 m s^{-1} and corresponds to an energy of 25 keV, and for α-particles the velocity is 0.44×10^7 m s^{-1} and the kinetic energy 400 keV. At lower energies the effective charge of the particle is reduced below z by electron pick-up and hence the rate of energy loss changes. Thus while an α-particle of energy 10 MeV has the same range as a proton of energy 2.5 MeV, to an accuracy of about 1 per cent, in accordance with the above rule, an α-particle of 800 keV has a range about 60 per cent greater than a proton of 200 keV. In the case of fission fragments, it is found that they already have about half their complement of electrons when they are emitted and the pick up of further electrons is a major effect along their track. This means that their effective charge is varying continuously along their trajectory.

If now we consider particles of the same kind and the same initial energy, and compare their ranges R_X and R_Y in two different absorbers, we have

$$\frac{R_X}{R_Y} = \frac{A_X Z_Y \rho_Y}{A_Y Z_X \rho_X},$$

providing the Z-dependence of $\bar{I}$ is negligible. For light materials $Z \simeq \frac{1}{2} A$ and hence

$$\frac{R_X}{R_Y} \simeq \frac{\rho_Y}{\rho_X}.$$

Thus the product $R_X \rho_X$ is approximately constant. It is therefore useful in comparing ranges in gases with, say, ranges in solids to compare the products of range and density, which will be of the same order of magnitude, rather than the linear ranges, which will differ by three orders of magnitude. The product will have the dimension of gm m^{-2}. It is customary to specify targets, windows, effective detector thicknesses, etc. in these units rather than in linear thicknesses.

The *stopping power* of an absorber is defined as $- dT/dx$ and is expressible in MeV cm^{-1}. It depends on the velocity of the incident particle.

The *relative stopping power* on the other hand is defined as the ratio of the amount of a standard substance (usually air at 15°C and atmospheric pressure) to the amount of the specified absorber in which there is the same energy loss. The amount may be measured as a distance, in which case we have the linear relative stopping power, or as a mass, which gives the mass stopping power, or as the number of electrons, which gives the relative stopping power per atomic electron.

We have now considered the factors in equation B.3 which depend on the properties of the absorber and on the static properties of the particle. The dependence of range on particle energy is contained in the integral. If we neglect the variation of the logarithmic denominator, then the range is expected to be proportional to V^4. This is in agreement with observed ranges for high initial particle velocities. However, if V_0 is reduced, the power falls below four and reaches about 1·5 at low velocities.

The energy lost by the particle in the absorber is expended in producing ionization and atomic excitations. In many applications it is important to know how much energy goes into each of these processes. Knowledge of this is largely empirical, and we restrict ourselves here to setting up the framework in which the empirical results can be contained.

The ionization can be divided into *primary* and *secondary*, the secondary ionization arising from the higher-energy electrons from the primary events, the so called δ-*rays*, which ionize in their own right as they stop in the absorber. The *total ionization* produced before the particle is stopped is distributed along the track and can be broken down into the *specific ionization*, which is the number of ion pairs per unit track length. The specific ionization rises as the particle velocity falls along the track, and it reaches a peak (6600 ion pairs per air-mm for α-particles, 2750 ion pairs per air-mm for protons) close to the end of the track. The total ionization for a given particle energy can be estimated from the empirical measurements of the average energy expended per ion pair produced. This quantity, which varies by less than a factor of two for different gases and for heavy particles of any velocity, is 35·5 eV for air, rising to 42·7 eV in helium and falling to 21·9 eV in xenon.

The processes of energy loss throughout the particle's travel and the processes of electron pick-up towards the end of the track involve individual random events. Thus we expect a statistical variation in the range of one particle with respect to another of the same initial energy. This is referred to as *range straggling*. For α-particles of 5·305 MeV, which have a mean range of 38·4 air-mm, the standard deviation of the actual ranges about the mean is 0·36 air-mm, i.e.

0·9 per cent. Protons show about twice the range straggling of α-particles of the same initial velocity and, as we saw above, roughly the same range.

B.2 Electrons

The problem of the energy loss and range of electrons is much more complicated than the same problem for heavy particles, for the following reasons.

(a) The electrons are very susceptible to scattering by nuclei and by orbital electrons. As a consequence the electron pursues a very tortuous path. We have therefore to distinguish between track length, as can be measured say in a cloud-chamber photograph of an electron track, and range, which is the maximum distance penetrated into an absorber as would be determined by placing foils in front of a detector.

(b) The bremstrahlung cross-section for electrons is appreciable and hence the energy loss by radiation is not negligible compared to the energy loss by ionization. When radiation loss takes place a large fraction of the total energy of the electron may be lost in one nuclear encounter.

(c) The electron mass is exactly equal to the orbital electron mass and hence it may lose any amount of energy up to its total energy in one collision with an orbital electron. These events, in which there is a high fractional loss of energy by radiation or electron collision, increase considerably the statistical fluctuations of energy loss and result in the range-straggling for electrons being very much greater than for heavy particles.

(d) The electron velocity at 1 MeV is 0·94 of the velocity of light, and hence a relativistic treatment is necessary. Here we shall only consider some empirical information about the absorption of electrons in foils.

For monokinetic electrons from, for example, an internally converted nuclear γ-ray line, the number of electrons transmitted through a foil falls off almost in direct proportion to the foil thickness until a small 'tail' is reached at large thicknesses. If this tail is ignored and the linear part of the curve produced to cut the thickness axis, then we get the extrapolated electron range R_0. This plays the role of the range R which in the case of heavy particles would be the foil thickness at which the transmission dropped abruptly from a steady value to zero.

Various empirical formulae have been suggested for different ranges of electron energies, with special reference to aluminium, which has been extensively used as a standard absorber in electron-range measurements. The expression

$$R_0 = 412 T_0^{(1·265 - 0·0954 \ln T_0)},$$

was found by Katz and Penfold (1952) to describe to the accuracy of a few per cent the experimental measurements in an energy range $10 \text{ keV} \leqslant T_0 \leqslant 2·5 \text{ MeV}$.

Nuclear β-rays are not monokinetic and their transmission through foils more closely follows an exponential curve than a straight line. Hence the extrapolation to find what, in this context, is termed the maximum range R_m, is less straightforward than for monokinetic electrons. Feather (1938) devised a method for arriving at a value of R_m by comparison with a standard β-spectrum. It is experimentally found, very conveniently, that R_m thus determined corresponds very well with R_0 for monokinetic electrons of energy equal to the maximum energy of the β-spectrum.

Appendix C
The Interaction of Gamma Rays with Matter

In passing through a material substance a γ-ray, in principle, may interact with (a) the system of orbital electrons in the absorber atoms, (b) the nuclei of the absorber atoms, and (c) the Coulomb field in the neighbourhood of the nucleus, arising from the electric charge of the protons.

In the energy range with which we are concerned, up to say 10 MeV, (b) can be neglected, the resonant absorption of γ-rays by the nucleus being a relatively unlikely process. It is found that there remain three main processes whose relative importances vary with the atomic number of the absorber and $h\nu$, the γ-ray energy. Before proceeding to consider each of these in turn we pause to define cross-section with respect to γ-ray processes.

C.1 Definition of cross-section for photons

Let us assume that we have a collimated beam of photons, cross-sectional area 1 cm^2, falling on an absorber of thickness dx, as shown in Figure 93. Let n photons fall on the absorber in a given time. In the absence of the absorber, all n of these photons, we assume, would have passed through the second collimator and been detected. The intervention of the absorber, let us assume, reduces the transmitted beam to $n - dn$, this attenuation being due to scattering and absorption

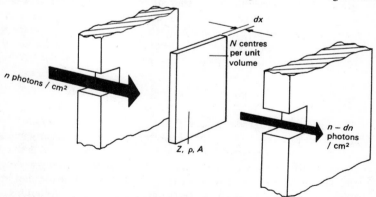

Figure 93 Collimated beam of photons passing through thin absorber

processes in the foil. If there are N centres per unit volume in the foil giving rise to these processes then we define the *total cross-section* σ by the equation

$$-\frac{dn}{n} = N\sigma \, dx. \hspace{3cm} \text{C.1}$$

If we can specify a subdivision of the scattering events, say by requiring a scattered photon to go off into a solid angle $d\Omega$ at an angle θ, then we can define on the same basis a *differential cross-section* $d\sigma$ in terms of $dn(\theta)$, the photons which interact *and* which result in a scattered photon in the specified direction.

A useful standard cross-section is the Thomson cross-section ϕ_0 for the scattering of a classical electromagnetic wave by a free electron. From classical theory

$$\phi_0 = \frac{8\pi}{3}\left(\frac{e^2}{m_0 c^2}\right)^2 = 6\cdot651 \times 10^{-25} \, \text{cm}^2.$$

C.2 The photoelectric effect

The γ-ray, in passing through an atom, may induce a transition in which a bound electron is raised to a level in the continuum of levels above the first ionization energy. The γ-ray then disappears and an electron with energy $h\nu - B$, where B is the electron binding energy, is produced. B will vary from shell to shell within the atom. If we consider the cross-section for this process as a function of $h\nu$, we find that, for values of $h\nu$ less than the binding energy of electrons in the atoms of the absorber, the cross-section is steadily falling but periodically jumps up discontinuously as the photon energy becomes large enough to release electrons from the next deeper shell. These discontinuities in the cross-section constitute the photoelectric *absorption edges*, which are well-known for X-ray absorption. They are equally important for low-energy γ-rays. For photon energies above the K-absorption edge (1·5 keV in aluminium; 88 keV in lead), 80 per cent of the photoelectrons are emitted from the K-shell.

Above the K-absorption edge, for the purposes of rough calculation, the atomic cross-section for the photoelectric process can be expressed as

$$\sigma_{\text{PE}} = \text{constant} \times \frac{Z^4}{(h\nu)^3}.$$

A knowledge that $\sigma_{\text{PE}} \approx \phi_0 = 6\cdot651 \times 10^{-25} \, \text{cm}^2$ for lead at 5 MeV then enables an estimate to be made of σ_{PE} for any absorber at any energy.

At low photon energies the photoelectrons tend to be emitted in the direction of the electric vector, i.e. at right angles to the γ-ray beam; at higher energies the electrons become directed forward. The most likely angle for photoelectrons from 100 keV photons is $50°$ to the beam and from 2·76 MeV photons is $10°$.

C.3   The Compton effect

A γ-ray, energy $h\nu$, may 'collide' with a free electron, the γ-ray scattering with reduced energy, $h\nu'$, and the electron being 'knocked on' with a kinetic energy T_e as in Figure 94. The application of linear-momentum and energy conservation to the collision leads to the result

$$h\nu' = \frac{m_0 c^2 h\nu}{m_0 c^2 + h\nu(1 - \cos\theta)}$$

$$= \frac{m_0 c^2}{1 - \cos\theta + 1/\alpha}, \quad \text{where} \quad \alpha = \frac{h\nu}{m_0 c^2}.$$

For back-scattered photons, i.e. where $\theta = 180°$,

$$h\nu' \rightarrow \frac{m_0 c^2}{2} = 0.25 \text{ MeV},$$

as $h\nu \rightarrow \infty$. For photons scattered through $90°$,

$$h\nu' \rightarrow h\nu,$$

for low-energy photons, and

$$h\nu' \rightarrow m_0 c^2 = 0.5 \text{ MeV},$$

as $h\nu \rightarrow \infty$.

The electron energy is given by

$$T_e = h\nu \frac{2\alpha \cos^2\phi}{(1 + \alpha)^2 - \alpha^2 \cos^2\phi}.$$

The maximum value of T_e occurs for $\phi = 0$ and

$$(T_e)_{\text{max}} = \frac{h\nu}{1 + 1/2\alpha}.$$

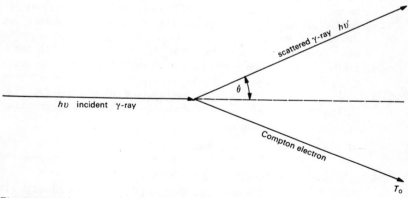

Figure 94   Kinematics of the Compton effect

For high-energy photons, therefore,

$$(T_e)_{max} = h\nu - \frac{m_0 c^2}{2},$$

in conformity with the above result for back-scattered photons.

The cross-section for the Compton process is given by the Klein–Nishina (1928) formula. The complete expression is very complicated. The following asymptotic expressions are frequently useful,

$$(\sigma_c)_e = \phi_0 (1 - 2\alpha + \tfrac{26}{5}\alpha^2 + \ldots \quad \text{for} \quad \alpha < 1,$$

$$(\sigma_c)_e = \frac{3}{8}\phi_0 \frac{\ln 2\alpha + \tfrac{1}{2}}{\alpha} \quad \text{for} \quad \alpha > 1,$$

where $(\sigma_c)_e$ is the cross-section per electron.

For energies well above the binding energies we assume that all atomic electrons are available for the process, and take the atomic cross-section to be given by $\sigma_c = Z(\sigma_c)_e$.

For 5 MeV γ-rays in lead $\quad \sigma_c \simeq 10\phi_0 = 6.651 \times 10^{-28} \text{ m}^2$.

C.4 Pair production

For γ-ray energies greater than $2m_0 c^2$ the process of pair production becomes energetically possible. An electron is raised across the gap between the filled negative-energy electron states and the states of positive kinetic energy, thus producing a free electron and leaving a vacancy in the negative-energy continuum, which vacancy behaves as a positron. Both electron and positron have linear momentum and it will be found impossible to balance the energy and momentum of the incoming γ-ray unless a third body takes part. Thus the process cannot take place in a vacuum but can proceed in the Coulomb field of a charged particle, momentum then being communicated to the particle. When the particle involved is a nucleus, the energy taken up by the nucleus is negligible and the threshold for the process lies very close to the energy necessary to produce the two electron rest masses, namely $2m_0 c^2 = 1 \cdot 02$ MeV. A less-likely process is for the pair production to take place in the field of an electron. In this event, at threshold energy in the laboratory system, all three electrons will be projected forward with the same energy and the threshold energy is then $4m_0 c^2 = 2 \cdot 04$ MeV.

The cross-section for pair production per atom can very roughly be taken to be

$$\sigma_p = \text{constant} \times Z^2 \ln h\nu.$$

In lead at 5 MeV, $\sigma_P \simeq \sigma_c \simeq 6 \cdot 651 \times 10^{-28} \text{ m}^2$.

C.5 Attenuation coefficients

In practice we are concerned with the transmission of γ-rays through a finite thickness of absorber. We recall, from equation C.1, that we can write

$$-\frac{dn}{n} = N\sigma\, dx.$$

The situation with respect to the variation of photon numbers with distance travelled through the absorber is seen to be exactly analogous to the situation with respect to the variation of the number of nuclei with time in a radioactive source, as discussed in Chapter 2. Thus the number of photons transmitted through a finite thickness x can be written as

$$n = n_0\, e^{-\mu x},$$

where n_0 is the number of incident photons and $\mu = N\sigma$. We call μ the linear attenuation coefficient and it has dimensions (length)$^{-1}$ In comparing the behaviour of gases with say solids then, as for heavy particles, it is convenient to work with absorbers measured in grammes per metre squared rather than in metres. We can rewrite the above equation as

$$n = n_0\, e^{-\rho x(\mu/\rho)}$$

$\mu/\rho = \mu_m$ is the mass-attenuation coefficient, customarily measured in centimetres squared per gramme and of the same order for all absorbers.

The values of μ_m for air and lead are plotted in Figure 95. These have been calculated from the cross-sections for the three processes and appear different because of the different Z-dependence of each of the processes.

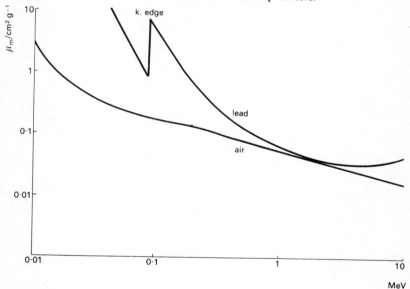

Figure 95 Mass-attenuation coefficients of γ-rays, as a function of energy, in lead and air

It should be noted that, for air, the photoelectric effect dominates for γ-ray energies below 50 keV, at which energy it gives way to the Compton effect, which then continues to dominate up to 20 MeV before pair production takes over. These energy boundaries in lead become 500 keV and 5 MeV, so that the region in which the Compton effect dominates is very much reduced in extent compared to the energy region in which the Compton effect is the dominant process in light materials.

Appendix D
General Atomic and Nuclear Constants

Velocity of light	c	$2{\cdot}997\,925\,0(10) \times 10^8$ m s^{-1}
Electron charge	e	$1{\cdot}602\,191\,7(70) \times 10^{-19}$ C
		$4{\cdot}803\,250 \times 10^{-10}$ e.s.u.
Planck constant	h	$6{\cdot}626\,196(50) \times 10^{-34}$ J s
Dirac constant	$\hbar$	$1{\cdot}054\,591\,9(80) \times 10^{-34}$ J s

Fine structure constant $\alpha \left(= \dfrac{e^2}{\hbar c} \right)$

$$\frac{1}{137{\cdot}036\,02(21)}$$

Boltzmann constant	k	$1{\cdot}380\,622(59) \times 10^{-23}$ J K^{-1}
		$8{\cdot}617\,084 \times 10^{-11}$ MeV K^{-1}
Mass of electron	m_e	$0{\cdot}511\,004\,1(16)$ MeV c^{-2}
		$\dfrac{1}{1836{\cdot}09} m_\text{p}$
		$5{\cdot}485\,97 \times 10^{-4}$ a.m.u.
Mass of proton	m_p	$1{\cdot}007\,276\,61(8)$ a.m.u.
		$938{\cdot}2592$ MeV c^{-2}
		$1836{\cdot}09\, m_\text{e}$
Mass of hydrogen atom	m_H	$1{\cdot}007\,826$ a.m.u.
Avagadro constant	N_A	$6{\cdot}022\,169 \times 10^{26}$ k mol^{-1}
Bohr magneton	μ_Bohr	$9{\cdot}274\,096(65) \times 10^{-28}$ J G^{-1}
		$0{\cdot}578\,838 \times 10^{-14}$ MeV G^{-1}
Nuclear magneton	μ_nucl	$5{\cdot}050\,951(50) \times 10^{-31}$ J G^{-1}
		$3{\cdot}152\,526 \times 10^{-18}$ MeV G^{-1}

Conversion factors

1 fm $= 10^{-15}$ m
1 barn $= 10^{-28}$ m^2
1 MeV $= 1\cdot602\,191\,7(70) \times 10^{-13}$ J
       $= 1\cdot073\,56 \times 10^{-3}$ a.m.u.
1 a.m.u $= 1\cdot660\,43 \times 10^{-24}$ g
       $= 931\cdot481\,2(52)$ MeV

(Based on B. N. Taylor, W. H. Parker and D. N. Langenberg, *Reviews of Modern Physics*, vol. 41, 1969, p. 375.)

(The numbers in parentheses are the standard deviation uncertainties in the last digits of the quoted value.)

Prefix	Symbol	Multiplying factor	Prefix	Symbol	Multiplying factor
Tera	T	10^{12}	*Deci	d	10^{-1}
Giga	G	10^{9}	*Centi	c	10^{-2}
Mega	M	10^{6}	Milli	m	10^{-3}
Kilo	k	10^{3}	Micro	μ	10^{-6}
*Hecto	h	10^{2}	Nano	n	10^{-9}
*Deca	da	10^{1}	Pico	p	10^{-12}
			Femto	f	10^{-15}
			Atto	a	10^{-18}

* Admitted in certain instances only.

Appendix E

E.1 List of the elements in order of charge number

1	H	hydrogen	36	Kr	krypton	71	Lu	lutetium
2	He	helium	37	Rb	rubidium	72	Hf	hafnium
3	Li	lithium	38	Sr	strontium	73	Ta	tantalum
4	Be	beryllium	39	Y	yttrium	74	W	tungsten
5	B	boron	40	Zr	zirconium	75	Re	rhenium
6	C	carbon	41	Nb	niobium	76	Os	osmium
7	N	nitrogen	42	Mo	molybdenum	77	Ir	iridium
8	O	oxygen	43	Tc	technetium	78	Pt	platinum
9	F	fluorine	44	Ru	ruthenium	79	Au	gold
10	Ne	neon	45	Rh	rhodium	80	Hg	mercury
11	Na	sodium	46	Pd	palladium	81	Tl	thallium
12	Mg	magnesium	47	Ag	silver	82	Pb	lead
13	Al	aluminium	48	Cd	cadmium	83	Bi	bismuth
14	Si	silicon	49	In	indium	84	Po	polonium
15	P	phosphorus	50	Sn	tin	85	At	astatine
16	S	sulphur	51	Sb	antimony	86	Rn	radon
17	Cl	chlorine	52	Te	tellurium	87	Fr	francium
18	A	argon	53	I	iodine	88	Ra	radium
19	K	potassium	54	Xe	xenon	89	Ac	actinium
20	Ca	calcium	55	Cs	caesium	90	Th	thorium
21	Sc	scandium	56	Ba	barium	91	Pa	protoactinium
22	Ti	titanium	57	La	lanthanum	92	U	uranium
23	V	vanadium	58	Ce	cerium	93	Np	neptunium
24	Cr	chromium	59	Pr	praseodymium	94	Pu	plutonium
25	Mn	manganese	60	Nd	neodymium	95	Am	americium
26	Fe	iron	61	Pm	promethium	96	Cm	curium
27	Co	cobalt	62	Sm	samarium	97	Bk	berkelium
28	Ni	nickel	63	Eu	europium	98	Cf	californium
29	Cu	copper	64	Gd	gadolinium	99	Es	einsteinium
30	Zn	zinc	65	Tb	terbium	100	Fm	fermium
31	Ga	gallium	66	Dy	dysprosium	101	Md	mendelevium
32	Ge	germanium	67	Ho	holmium	102	No	nobelium
33	As	arsenic	68	Er	erbium	103	Lw	lawrencium
34	Se	selenium	69	Tm	thulium	104	Ku	kurchatovium
35	Br	bromine	70	Yb	ytterbium			

E.2 List of the elements in alphabetical order

Actinium	Ac	89	Kurchatovium	Ku	104	Tin	Sn	50
Aluminium	Al	13	Lanthanum	La	57	Titanium	Ti	22
Americium	Am	95	Lawrencium	Lw	103	Tungsten	W	74
Antimony	Sb	51	Lead	Pb	82	Uranium	U	92
Argon	A	18	Lithium	Li	3	Vanadium	V	23
Arsenic	As	33	Lutetium	Lu	71	Xenon	Xe	54
Astatine	At	85	Magnesium	Mg	12	Ytterbium	Yb	70
Barium	Ba	56	Manganese	Mn	25	Yttrium	Y	39
Berkelium	Bk	97	Mendelevium	Md	101	Zinc	Zn	30
Beryllium	Be	4	Mercury	Hg	80	Zirconium	Zr	40
Bismuth	Bi	83	Molybdenum	Mo	42			
Boron	B	5	Neodymium	Nd	60			
Bromine	Br	35	Neon	Ne	10			
Cadmium	Cd	48	Neptunium	Np	93			
Caesium	Cs	55	Nickel	Ni	28			
Calcium	Ca	20	Niobium	Nb	41			
Californium	Cf	98	Nitrogen	N	7			
Carbon	C	6	Nobelium	No	102			
Cerium	Ce	58	Osmium	Os	76			
Chlorine	Cl	17	Oxygen	O	8			
Chromium	Cr	24	Praseodymium	Pr	59			
Cobalt	Co	27	Promethium	Pm	61			
Copper	Cu	29	Protoactinium	Pa	91			
Curium	Cm	96	Radium	Ra	88			
Dysprosium	Dy	66	Radon	Rn	86			
Einsteinium	Es	99	Rhenium	Re	75			
Erbium	Er	68	Rhodium	Rh	45			
Europium	Eu	63	Rubidium	Rb	37			
Fermium	Fm	100	Ruthenium	Ru	44			
Fluorine	F	9	Samarium	Sm	62			
Francium	Fr	87	Scandium	Sc	21			
Gadolinium	Gd	64	Selenium	Se	34			
Gallium	Ga	31	Silicon	Si	14			
Germanium	Ge	32	Silver	Ag	47			
Gold	Au	79	Sodium	Na	11			
Hafnium	Hf	72	Strontium	Sr	38			
Helium	He	2	Sulphur	S	16			
Holmium	Ho	67	Tantalum	Ta	73			
Hydrogen	H	1	Technetium	Tc	43			
Indium	In	49	Tellurium	Te	52			
Iodine	I	53	Terbium	Tb	65			
Iridium	Ir	77	Thallium	Tl	81			
Iron	Fe	26	Thorium	Th	90			
Krypton	Kr	36	Thulium	Tm	69			

Further Reading

General reading in the field of nuclear physics

W. E. BURCHAM, *An Introduction to Nuclear Physics*, Longman, 1965.
   This is an account at the same level of difficulty as the present text, and
covers approximately the same material but with a different emphasis.
C. M. H. SMITH, *A Textbook of Nuclear Physics*, Pergamon, 1965, and
R. M. EISBERG, *Fundamentals of Modern Physics*, Wiley, 1961.
   Written at the same level of difficulty as the present text, these two books
give a full account of the theoretical background. The former also contains
detailed accounts of the most important experimental developments.
E. SEGRÈ, *Nuclei and Particles*, Benjamin, 1964, and E. B. PAUL, *Nuclear and
Particle Physics*, North Holland, 1970.
   Written at a somewhat higher level of difficulty than the present text,
these two books give a full account of nuclear physics, and relate it to
elementary-particle physics.
An advanced authoritative text on the experimental aspects of nuclear
physics is R. D. EVANS, *The Atomic Nucleus*, McGraw-Hill, 1955. A
corresponding text on the theoretical aspects is J. M. BLATT and
V. F. WEISSKOPF, *Theoretical Nuclear Physics*, Wiley, 1952. Detailed
information on particular nuclei is to be found in C. M. LEDERER,
J. M. HOLLANDER and I. PERLMAN, *Table of Isotopes*, Wiley, 1967, whilst
information which will be found useful in quantitative planning and interpretation
of experiments is collected in A. H. WAPSTRA, G. J. NIJGH and R. VAN LIESHOUT.
Nuclear Spectroscopy Tables, North Holland, 1959.

Further reading for each chapter follows.

Chapter 1

J. J. THOMSON and G. P. THOMSON, *Conduction of Electricity through Gases*,
   Cambridge University Press, 1928.
J. A. CROWTHER, *Ions, Electrons and Ionising Radiations*, Arnold, 1944.
G. FEINBERG, 'Ordinary matter', *Scientific American*, May 1967.
G. P. THOMSON, *The Atom,* Oxford University Press, 1956.

Chapter 2

L. BADASH, 'How the "newer alchemy" was received', *Scientific American*,
   August 1966.

Chapter 3

J. D. STRANATHAN, *The Particles of Modern Physics*, Blakiston, 1946.

E. RUTHERFORD, J. CHADWICK and C. D. ELLIS, *Radiations from Radioactive Substances*, Cambridge University Press, 1950.

E. N. da C. ANDRADE, 'The birth of the nuclear atom', *Scientific American*, November 1956.

Chapter 4

S. B. TREIMAN, 'The weak interactions', *Scientific American*, March 1959.

P. MORRISON, 'The neutrino', *Scientific American*, January 1956.

F. REINES and J. P. F. SELLSCHOP, 'Neutrinos from the Atmosphere and Beyond', *Scientific American*, February 1966.

G. M. LEWIS, *Neutrinos*, Wykeham, 1970.

Chapter 5

A. E. S. GREEN, *Nuclear Physics*, McGraw-Hill, 1955.

Chapter 6

J. HAMILTON, 'Nuclear structure', *Endeavour*, No. 19, 1960. This article with some others relating to shell-model topics is reproduced in: *Nuclear Structure – Selected Reprints*, American Institute of Physics, 1965.

Chapter 7

G. HERZBERG, *Atomic Spectra and Atomic Structure*, Dover, 1944.

R. B. LEIGHTON, *Principles of Modern Physics*, McGraw-Hill, 1959.

H. KOPFERMANN, *Nuclear Moments*, (English Version) Academic Press, 1958.

Chapter 8

G. HERZBERG, *Atomic Spectra and Atomic Structure*, Dover, 1944.

H. KOPFERMANN, *Nuclear Moments*, Academic Press, 1958.

N. F. RAMSEY, *Molecular Beams*, Clarendon Press, 1956.

F. K. RITCHMYER and E. H. KENNARD, *Introduction to Modern Physics*, McGraw-Hill, 1954.

R. B. LEIGHTON, *Principles of Modern Physics*, McGraw-Hill, 1959.

O. R. FRISCH, 'Molecular beams, *Scientific American*, May 1965.

Chapter 10

Experimental Nuclear Physics, Volume 3, Wiley, 1959.

C. E. CROUTHAMEL, *Applied Gamma-Ray Spectrometry*, Pergamon, 1960.

G. DEARNALAY and D. C. NORTHROP, *Semiconductor Counters for Nuclear Radiations*, Wiley, 1963.

Chapter 11

L. WILETS, *Theories of Nuclear Fission*, Clarendon Press, 1964.

R. B. LEACHMAN, 'Nuclear Fission', *Scientific American*, August 1965.

Acknowledgements

Acknowledgement is due to the publishers for permission to use figures and tables from the following books.

Figure 1 Knolls Atomic Power Laboratory, Schenectady, New York (operated by the General Electric Company for the USAEC Naval Reactor Branch. A detail from their *Chart of the Nuclides.*)

Figure 13 American Institute of Physics, from J. L. Lawson and J. M. Cook, *Physical Review*, vol. 57, 1940, p. 994.

Figure 42 Academic Press Inc., from H. Kopfermann, *Nuclear Moments.*

Figure 56 Les Presses de L'Université de Montreal, from H. Noma, T. Shibata and Y. Yoshizawa, *Contributions to International Conference on Properties of Nuclear States*, Montreal, 1969.

Figures 60 and 61 North Holland Publishing Co., from E. B. Paul, *Nuclear and Particle Physics.*

Figure 64 Academic Press Inc. from *Nuclear Data Sheets* (NRC – 61; 5, 6, p. 185).

Figures 69 and 70 North Holland Publishing Co., from A. H. Wapstra, G. J. Nijgh and R. Van Lieshout, *Nuclear Spectroscopy Tables.*

Figures 71 and 72 McGraw-Hill Book Company, from R. D. Evans, *The Atomic Nucleus.*

Figure 73 Pergamon Press, from C. E. Crouthamel, *Applied Gamma Ray Spectroscopy.*

Figure 77 North Holland Publishing Co., from an article by R. L. Mossbauer in α, β, γ *Spectroscopy*, edited by K. Siegbahn.

Figures 78 and 79 Academic Press Inc., from an article by G. H. Fuller and V. W. Cohen in *Nuclear Data Tables*, A vol. 5, 1969.

Figure 84 University of Chicago Press, from A. M. Weinberg and E. Wigner, *The Physical Theory of Neutron Chain Reactors.*

Figures 86, 89 and 90 International Atomic Energy Agency, from V. M. Strutinsky and H. C. Pauli, *Proceedings of Second IAEA Symposium on Physics and Chemistry of Fission*, 1969.

Figures 87(b) and 88 From W. J. Swiatecki, *Annals of Physics*, vol. 19, 1962, p. 67. Oxford University Press, from L. Wilets, *Theories of Nuclear Fission.*

Index

Chapter 12, added in 1984, is not indexed